Chemical Conservation of Paper-Based Cultural Heritage

Chemical Conservation of Paper-Based Cultural Heritage

Guest Editors

Yueer Yan
Yi Tang
Yuliang Yang

Basel • Beijing • Wuhan • Barcelona • Belgrade • Novi Sad • Cluj • Manchester

Guest Editors

Yueer Yan
Institute for Preservation and Conservation of Chinese Ancient Books
Fudan University
Shanghai
China

Yi Tang
Department of Chemistry
Fudan University
Shanghai
China

Yuliang Yang
Department of Macromolecular Science
Fudan University
Shanghai
China

Editorial Office
MDPI AG
Grosspeteranlage 5
4052 Basel, Switzerland

This is a reprint of the Special Issue, published open access by the journal *Molecules* (ISSN 1420-3049), freely accessible at: www.mdpi.com/journal/molecules/special_issues/U1S89MW5JN.

For citation purposes, cite each article independently as indicated on the article page online and using the guide below:

Lastname, A.A.; Lastname, B.B. Article Title. *Journal Name* **Year**, *Volume Number*, Page Range.

ISBN 978-3-7258-3626-0 (Hbk)
ISBN 978-3-7258-3625-3 (PDF)
https://doi.org/10.3390/books978-3-7258-3625-3

© 2025 by the authors. Articles in this book are Open Access and distributed under the Creative Commons Attribution (CC BY) license. The book as a whole is distributed by MDPI under the terms and conditions of the Creative Commons Attribution-NonCommercial-NoDerivs (CC BY-NC-ND) license (https://creativecommons.org/licenses/by-nc-nd/4.0/).

Contents

Yueer Yan, Yi Tang and Yuliang Yang
Chemical Conservation of Paper-Based Cultural Heritage
Reprinted from: *Molecules* **2024**, *30*, 122, https://doi.org/10.3390/molecules30010122 1

Zhiyou Han, Keiko Kida, Kyoko Saito Katsumata, Masaki Handa and Masamitsu Inaba
Effect of Hemicellulose on the Wet Tensile Strength of *Kozo* Paper
Reprinted from: *Molecules* **2023**, *28*, 6996, https://doi.org/10.3390/molecules28196996 7

Zirui Zhu, Kai Zhang, Yu Xue, Zhongming Liu, Yujie Wang and Yanli Zhang et al.
Research on the Structure and Properties of Traditional Handmade Bamboo Paper During the Aging Process
Reprinted from: *Molecules* **2024**, *29*, 5741, https://doi.org/10.3390/molecules29235741 20

Wenjie Zhang, Shan Wang and Hong Guo
Influence of Relative Humidity on the Mechanical Properties of Palm Leaf Manuscripts: Short-Term Effects and Long-Term Aging
Reprinted from: *Molecules* **2024**, *29*, 5644, https://doi.org/10.3390/molecules29235644 38

Haibo Zhao, Hongbin Zhang, Qiang Xu, Hongdong Zhang and Yuliang Yang
Thermal, Rheological, Structural and Adhesive Properties of Wheat Starch Gels with Different Potassium Alum Contents
Reprinted from: *Molecules* **2023**, *28*, 6670, https://doi.org/10.3390/molecules28186670 54

Chunfang Wu, Yangyang Liu, Yanxiao Hu, Ming Ding, Xiang Cui and Yixin Liu et al.
An Investigation into the Performance and Mechanisms of Soymilk-Sized Handmade Xuan Paper at Different Concentrations of Soymilk
Reprinted from: *Molecules* **2023**, *28*, 6791, https://doi.org/10.3390/molecules28196791 70

Gele Teri, Cong Cheng, Kezhu Han, Dan Huang, Jing Li and Yujia Luo et al.
Study on the Properties of FEVE Modified with Ag_2O/OH-MWCNTS Nanocomposites for Use as Adhesives for Wooden Heritage Objects
Reprinted from: *Molecules* **2024**, *29*, 1365, https://doi.org/10.3390/molecules29061365 85

Catalin Croitoru and Ionut Claudiu Roata
Ionic Liquids as Reconditioning Agents for Paper Artifacts
Reprinted from: *Molecules* **2024**, *29*, 963, https://doi.org/10.3390/molecules29050963 98

Mengruo Wu, Le Mu, Zhiyue Zhang, Xiangna Han, Hong Guo and Liuyang Han
Anti-Cracking TEOS-Based Hybrid Materials as Reinforcement Agents for Paper Relics
Reprinted from: *Molecules* **2024**, *29*, 1834, https://doi.org/10.3390/molecules29081834 114

Yi Wang, Zirui Zhu, Jinhua Wang, Peng Liu, Xingxiang Ji and Hongbin Zhang et al.
Facile Synthesis of Low-Dimensional and Mild-Alkaline Magnesium Carbonate Hydrate for Safe Multiple Protection of Paper Relics
Reprinted from: *Molecules* **2024**, *29*, 4921, https://doi.org/10.3390/molecules29204921 129

Hongyan Mou, Ting Wu, Xingxiang Ji, Hongjie Zhang, Xiao Wu and Huiming Fan
Multi-Functional Repair and Long-Term Preservation of Paper Relics by Nano-MgO with Aminosilaned Bacterial Cellulose
Reprinted from: *Molecules* **2024**, *29*, 3959, https://doi.org/10.3390/molecules29163959 143

Editorial

Chemical Conservation of Paper-Based Cultural Heritage

Yueer Yan [1,*], Yi Tang [2,*] and Yuliang Yang [1,3,*]

1. Institute for Preservation and Conservation of Chinese Ancient Books, Fudan University Library, Fudan University, Shanghai 200433, China
2. Department of Chemistry, Fudan University, Shanghai 200433, China
3. Department of Macromolecular Science, Fudan University, Shanghai 200433, China
* Correspondence: yueeryan@fudan.edu.cn (Y.Y.); yitang@fudan.edu.cn (Y.T.); yulyang1952@163.com (Y.Y.)

Received: 26 December 2024
Accepted: 28 December 2024
Published: 31 December 2024

Citation: Yan, Y.; Tang, Y.; Yang, Y. Chemical Conservation of Paper-Based Cultural Heritage. *Molecules* **2025**, *30*, 122. https://doi.org/10.3390/molecules30010122

Copyright: © 2024 by the authors. Licensee MDPI, Basel, Switzerland. This article is an open access article distributed under the terms and conditions of the Creative Commons Attribution (CC BY) license (https://creativecommons.org/licenses/by/4.0/).

Paper-based cultural heritages, represented by ancient books, archives, calligraphy, and paintings, recorded the development of human civilization. However, they are inevitably deteriorated during their long-term preservation, hence requiring intensive care to protect and inherit them. Chemists can play crucial roles in this mission, considering the interdisciplinary nature of the field of heritage science. The relationship between heritage science and applied chemistry is particularly close when facing the problems of paper aging, where intrinsic and extrinsic factors are entangled to form convoluted degradation mechanisms. Since the public awareness of cultural heritage protection is increasing, an ever-growing effort has been devoted to the preservation and conservation of paper artifacts.

This Special Issue successfully collected the important contributions in the field of heritage conservation. Contributions include studies on the properties and degradation mechanisms of carriers of cultural heritages, the characteristics of adhesives commonly used in mounting and restoration, and the various materials employed in the cleaning, deacidification, and reinforcement of paper-based artifacts. By carrying out comprehensive research on paper-based cultural heritage, this Special Issue aims to provide a deeper understanding of the mechanisms involved in the degradation and preservation of paper artifacts and offers practice insights into more efficient and sustainable conservation methods.

Paper has been focused on its unique physical and chemical properties when used as the carrier of paper-based cultural heritage. Japanese *kozo* paper is commonly used as the first back lining paper of hanging scrolls, which plays a crucial role in supporting the main paper with a painting or a calligraphy on it. This lining paper is often dyed and adhered during the mounting process and, therefore, a high wet tensile strength is essential. Han et al. investigated the effects of hemicellulose (especially glucuronoxylan) on the wet tensile strength of *kozo* paper produced in different cooking conditions (contribution 1). The wet tensile strength of paper was evaluated using the Finch device, and the content of glucuronoxylan in fiber was analyzed by a gas chromatographic method. By analyzing the glucuronoxylan content and wet tensile strength of various *kozo* paper samples, the authors revealed that papers with higher glucuronoxylan content exhibited stronger wet tensile strength. Then, they studied the effects of cooking time and the type and concentration of alkali on the glucuronoxylan content of paper under varying cooking conditions. The results indicated that the cooking time had minimal impact, while using alkali with weaker alkalinity and lower concentration resulted in higher glucuronoxylan retention within the fibers and stronger wet strength of *kozo* paper. Furthermore, they added the extracted glucuronoxylan into the *kozo* pulp under identical cooking conditions. The as-prepared papers showed a proportional increase in wet tensile strength with increasing glucuronoxylan content, indicating that glucuronoxylan itself contributes significantly to the wet tensile strength of *kozo* paper. This work provides a guideline for the production of *kozo* paper with an expected wet tensile strength and contributes to the preservation and conservation of cultural heritage with good paper quality.

The existence of hemicellulose not only affects paper's wet tensile strength, but also influences the long-term stability of paper. Zhu et al. studied the structure and properties of Chinese bamboo paper during the aging process and focused on the influence of hemicellulose and lignin on paper durability (contribution 2). It was found that the mechanical properties of bamboo paper with high lignin and hemicellulose content exhibited three stages during dry-heat aging: initial plateau, rapid decline, and second plateau as the degree of polymerization (DP) of cellulose decreased. The contents of cellulose and hemicellulose decreased while lignin remained relatively stable. By analyzing the changes in microstructures during the aging process, such as hydrogen bonds, hornification, pore structure, and crystallinity, the aging mechanisms of bamboo paper were proposed. The decrease in DP of cellulose led to fiber embrittlement, but the enhancement of hydrogen bonds and the hornification process could partially compensate for this effect, thereby maintaining the mechanical properties of paper during the initial plateau stage. When the DP of cellulose fell below the critical value, the deep cleavage of cellulose chains and the release of a large number of degradation products caused a decrease in cellulose content and an increase in paper acidity, resulting in a sharp decline in paper mechanical properties. The existence of hemicellulose and lignin might contribute to the stability of paper structure by inhibiting cellulose aggregation, thus, forming a second plateau stage. This research not only improves the comprehension of the relationship between the structural properties and degradation mechanisms of handmade papers but also provides a scientific foundation for the production of handmade paper with better durability.

Similarly to paper, palm leaves are used as writing and recording materials, derived from plant resources. Palm leaves served as a popular literary medium in South and Southeast Asia before the widespread use of paper, which played an invaluable role in the development of human civilization. Studies on the mechanical properties of palm leaf manuscripts under different environmental conditions are of great significance to understand the material characteristics, aging mechanisms, and preventive conservation methods of these manuscripts. Zhang et al. investigated the short-term and long-term effects of relative humidity (RH) on the mechanical properties of palm leaf manuscripts (contribution 3). In their experiments, dynamic vapor sorption (DVS) was used to measure the hygroscopic behavior of palm leaves in response to different humidity conditions, and a thermomechanical analyzer–modular humidity generator (TMA-MHG) was used to analyze the mechanical behavior (bending strength and bending modulus) of the samples at various humidity levels. The short-term study showed that the bending strength of the palm leaf samples decreased significantly with the increase in humidity, while the bending modulus firstly increased and then decreased. A stable RH could slow down the variations in the mechanical properties, thereby preventing the degradation of palm leaf materials. The long-term results indicated that the prolonged exposure to either extremely dry or humid environments could seriously affect the mechanical properties of palm leaves, which was detrimental to their preservation. Samples aged at 50% RH had no visible damage, and the mechanical properties and chemical structures were well preserved, suggesting that 50% RH should be an optimal humidity condition for preserving palm leaf manuscripts. This study provides a comprehensive understanding of the mechanical behaviors of palm leaf manuscripts and supports the development of long-term preservation and preventive conservation of these priceless manuscripts.

For paper-based cultural heritage, the use of adhesives plays a crucial role, especially in artwork mounting and paper sizing. Wheat starch has been a common adhesive material used in the mounting of calligraphies and paintings for over a millennium. Since the naturally derived starch gel has certain limitations, such as relatively low viscosity and easy retrogradation, Chinese restorers discovered that the combination of wheat starch with potassium alum (PA) offered promising improvements in the adhesive strength and stability of starch gel. Zhao et al. investigated the effects of varying PA contents on the thermal, rheological, structural, and adhesive properties of wheat starch gels (contribution 4). The results indicated that the incorporation of PA elevated the gelatinization temperature and enthalpy of the starch gels, with a maximum leached amylose and swelling power observed at a specific wheat starch to PA weight ratio of 100:6. Rheological measurements and scanning electron microscopy (SEM) observations corroborated this phenomenon. Fourier transform infrared spectroscopy (FTIR) analysis showed an enhanced short-range molecular order in the gels with PA. Mechanical experiments revealed that the binding strength of the starch gels increased with higher PA concentrations and decreased significantly after the aging process. Therefore, it is advisable to strictly control the concentration of PA in starch gels to balance the enhanced rheological and adhesive properties with the minimization of negative impacts on the degradation and color change of paper relics. This study provides a comprehensive understanding of the effects of PA on starch gels and offers practical insights for the restoration of valuable cultural artifacts.

Starch gel and alum gelatin are also widely used as natural sizing agents for paper. However, natural starch has a poor affinity with fibers and a tendency to detach, and gelatin alum poses a threat to the durability of paper. Hence, studies on efficient and harmless agents for paper sizing are still required in the conservation of cultural heritage. Wu et al. investigated the use of soymilk as a sizing agent for Xuan paper, focusing on its impact on paper properties and long-term stability (contribution 5). They prepared different concentrations of soymilk and characterized its pH, viscosity, particle size, and zeta potential. The soymilk was applied to Xuan paper, and the soymilk-sized paper had improved hydrophobicity, enhanced mechanical properties, and unique chromaticity compared to unsized paper. These characteristics are attributed to the papillae on the surface of Xuan paper, the folding of the soy protein, and the hydrogen bond interactions between the soy protein and paper fibers. The soymilk-sized paper also showed a higher initial pH, which declined more slowly than alum gelatin-sized paper, suggesting better acid resistance. After accelerated aging, the mechanical strength and color fidelity of paper deteriorated, and the expected lifespans of paper decreased with the increasing soymilk concentration. Compared with paper sized with alum gelatin, the soymilk-sized paper showed better hydrophobicity and durability. This study highlights the potential benefits and challenges of using soymilk as a paper sizing agent and aids the selection of proper materials for the mounting and conservation of paper artifacts.

Additionally, research on adhesives has been expanded to the protection of wooden heritage objects. Wood composed of cellulose, hemicellulose, and lignin is a natural material whose physical and chemical properties change during the preservation process, leading to varying degrees of degradation. Traditional adhesives used in conservation of wood artifacts have their limitations, particularly in terms of low adhesion, durability, and antimicrobial properties. Teri et al. investigated the synthesis and application of a novel adhesive for preserving wooden heritage objects. The adhesive is prepared by modifying fluorinated ethylene vinyl ether (FEVE) resin with Ag_2O-decorated hydroxyl multiwall carbon nanotubes (Ag_2O/OH-MWCNTS, MA) composites, and the prepared adhesive was named MAF (contribution 6). The resulting MAF adhesive showed enhanced thermal stability, higher viscosity, and effective bacteriostatic property compared to pure

FEVE resin. The viscosity and adhesion strength of MAF increased with the increase in MA concentration, indicating that Ag_2O nanoparticles acted as physical crosslinking agents within the FEVE resin. Compared with pure FEVE resin, the MAF adhesive had a lower mass loss rate after treatment at 800 °C and exhibited better thermal stability. The antibacterial activity of MAF is directly related to the concentration of Ag_2O/OH-MWCNTS, with the significant inhibitory effect on *Trichoderma, Aspergillus niger*, and *white rot fungi*. The MAF adhesive shows promising prospects for use in wooden heritage artifacts, offering a novel approach for the rapid, environmentally friendly, and efficient preparation of composite adhesives with superior adhesive properties.

The restoration of paper artifacts is of great importance to ensure their longevity. However, traditional restoration methods are time-consuming and always contain multistep processes, which highlights the need for alternative methods that can provide a more effective, gentle, and sustainable approach to protect cultural heritages. One potential solution that has emerged in recent years is the use of ionic liquids. Ionic liquids are composed entirely of ions, which gives them unique properties and makes them suitable for paper cleaning and antimicrobial purposes. Croitoru et al. investigated the application of ionic liquids for restoring aged paper artifacts, focusing on their ability to clean, deacidify, and stabilize against ultraviolet (UV) radiation (contribution 7). The treated paper exhibited significant color enhancements, indicating a good cleaning ability of ionic liquids. Moreover, the selected ionic liquids displayed pronounced deacidification capabilities, leading to a neutral pH of paper. They also acted as potential UV stabilizers to minimize the yellowing and discoloration of paper caused by UV radiation. All these results demonstrate the potential use of ionic liquids as a novel and effective approach for the restoration of paper artifacts.

The deterioration of paper always causes its embrittlement and the loss of mechanical strength. Organosilane materials have emerged as promising candidates for paper reinforcement due to their dual advantages of organic and inorganic properties. Wu et al. introduced dodecyltrimethoxysilane (DTMS) into tetraethoxysilane (TEOS) to prepare the hybrid materials as reinforcement agents for paper relics (contribution 8). By incorporating DTMS with flexible long chains, they aimed to weaken the capillary forces and minimize the shrinkage during the sol–gel process of TEOS. The gel formed from the DTMS/TEOS hybrid material was transparent and crack-free, featuring a dense microstructure with excellent thermal stability. When applied to bamboo paper, the hybrid material combined with the paper fibers through the sol–gel process and polymerized into a network structure that enveloped the paper surface or penetrated between the fibers. The surface of the treated paper displayed excellent hydrophobic properties, with no significant changes in appearance, color, or air permeability. The tensile strengths of the treated bamboo paper improved significantly, indicating the efficient reinforcement of paper by DTMS/TEOS. The developed hybrid material demonstrates a positive reinforcement and preservation effect on paper-based cultural relics. The advantages of ease of preparation, low cost, and wide accessibility make DTMS/TEOS a promising choice for future applications in paper conservation.

In addition to brittleness, the aging and degradation of paper also lead to other problems such as yellowing, discoloration, and acidification. Hence, new restoration materials are emerging for the multifunctional protection of paper-based cultural relics. Wang et al. synthesized low-dimensional and mildly alkaline magnesium carbonate hydrates (MCHs) for the deacidification, strengthening, and flame retardance of paper relics (contribution 9). The one-dimensional (1D) whisker and two-dimensional (2D) nanosheet MCHs are prepared through a simple supersaturation control method without the need for organic additives. The results indicated that MCHs with mild alkalinity exhibited good safety, showing minimal color change to the paper and little effect on alkali-sensitive pigments. Both 1D and 2D MCHs effectively neutralized the acidity in the paper and maintained a long-term alkaline reserve. Due to the structural properties of two kinds of MCHs, 1D MCHs showed better performance than 2D MCHs in enhancing the mechanical strength of paper and providing a flame-retardant effect. The ability to control the microstructure of MCHs allows for the design of materials with specific properties for different applications. This study also gives a valuable insight into the development of new conservation materials and expands the understanding of functional deacidification and protection mechanisms.

The use of composite materials provides more functions and better performance for the conservation of aged paper. Nanocellulose with natural paper compatibility has become a promising nanomaterial for paper repairing. The combination of nanocellulose with alkaline nanoparticles has been widely applied for the simultaneous reinforcement and deacidification of paper. Mou et al. prepared magnesium oxide (MgO) nanoparticles carried by aminosilaned bacterial cellulose (KH550-BC) for the restoration of aged paper (contribution 10). The existence of MgO in KH550-BC enhanced the paper pH and provided an alkaline reserve to counteract future acidification, while KH550-BC improved the mechanical properties of paper, including tear index, tensile index, and folding endurance. The hydrogen bond interactions between KH550-BC and paper fibers as well as the formed network structure of BC enhanced the paper strength. The hydrophobicity of paper is maintained after treatment, preventing damage from moisture and humidity. Artificial aging tests demonstrated that KH550-BC/MgO-treated paper exhibited superior resistance to acidification, mechanical degradation, and yellowing compared to untreated paper. The results of this work indicate that KH550-BC/MgO is an effective deacidification and reinforcement material for paper-based cultural heritage.

In conclusion, research about the chemical conservation of paper-based artifacts, especially focusing on the properties and applications of carriers, adhesives, and restoration materials, holds significant importance in the field of heritage science. Through continuous technological innovation and interdisciplinary integration, we are poised to discover more effective conservation solutions for the inheritance and protection of precious cultural heritage.

Conflicts of Interest: The authors declare no conflicts of interest.

List of Contributions

1. Han, Z.; Kida, K.; Katsumata, K.S.; Handa, M.; Inaba, M. Effect of Hemicellulose on the Wet Tensile Strength of *Kozo* Paper. *Molecules* **2023**, *28*, 6996.
2. Zhu, Z.; Zhang, K.; Xue, Y.; Liu, Z.; Wang, Y.; Zhang, Y.; Liu, P.; Ji, X. Research on the Structure and Properties of Traditional Handmade Bamboo Paper During the Aging Process. *Molecules* **2024**, *29*, 5741.
3. Zhang, W.; Wang, S.; Guo, H. Influence of Relative Humidity on the Mechanical Properties of Palm Leaf Manuscripts: Short-Term Effects and Long-Term Aging. *Molecules* **2024**, *29*, 5644.
4. Zhao, H.; Zhang, H.; Xu, Q.; Zhang, H.; Yang, Y. Thermal, Rheological, Structural and Adhesive Properties of Wheat Starch Gels with Different Potassium Alum Contents. *Molecules* **2023**, *28*, 6670.
5. Wu, C.; Liu, Y.; Hu, Y.; Ding, M.; Cui, X; Liu, Y.; Liu, P.; Zhang, H.; Yang, Y.; Zhang, H. An Investigation into the Performance and Mechanisms of Soymilk-Sized Handmade Xuan Paper at Different Concentrations of Soymilk. *Molecules* **2023**, *28*, 6791.
6. Teri, G.; Cheng, C.; Han, K.; Huang, D.; Li, J.; Luo, Y.; Fu, P.; Li, Y. Study on the Properties of FEVE Modified with Ag_2O/OH-MWCNTS Nanocomposites for Use as Adhesives for Wooden Heritage Objects. *Molecules* **2024**, *29*, 1365.
7. Croitoru, C.; Roata, I.C. Ionic Liquids as Reconditioning Agents for Paper Artifacts. *Molecules* **2024**, *29*, 963.

8. Wu, M.; Mu, L.; Zhang, Z.; Han, X.; Guo, H.; Han, L. Anti-Cracking TEOS-Based Hybrid Materials as Reinforcement Agents for Paper Relics. *Molecules* **2024**, *29*, 1834.
9. Wang, Y.; Zhu, Z.; Wang, J.; Liu, P.; Ji, X.; Zhang, H.; Tang, Y. Facile Synthesis of Low-Dimensional and Mild-Alkaline Magnesium Carbonate Hydrate for Safe Multiple Protection of Paper Relics. *Molecules* **2024**, *29*, 4921.
10. Mou, H.; Wu, T.; Ji, X.; Zhang, H.; Wu, X.; Fan, H. Multi-Functional Repair and Long-Term Preservation of Paper Relics by Nano-MgO with Aminosilaned Bacterial Cellulose. *Molecules* **2024**, *29*, 3959.

Disclaimer/Publisher's Note: The statements, opinions and data contained in all publications are solely those of the individual author(s) and contributor(s) and not of MDPI and/or the editor(s). MDPI and/or the editor(s) disclaim responsibility for any injury to people or property resulting from any ideas, methods, instructions or products referred to in the content.

Article

Effect of Hemicellulose on the Wet Tensile Strength of *Kozo* Paper

Zhiyou Han [1,*], Keiko Kida [2], Kyoko Saito Katsumata [2,3], Masaki Handa [4] and Masamitsu Inaba [2,*]

1. Conservation Standards Research Institute, The Palace Museum, Beijing 100009, China
2. Conservation Science Laboratory, Graduate School of Conservation, Tokyo University of the Arts, Tokyo 110-8714, Japan; kida.keiko@fa.geidai.ac.jp (K.K.); katsumata.kyoko@fa.geidai.ac.jp (K.S.K.)
3. Graduate School of Agricultural and Life Sciences, Faculty of Agriculture, University of Tokyo, Tokyo 113-8654, Japan
4. Handa Kyuseido Co., Ltd., Tokyo 151-0064, Japan; masaki-handa@kyuseido.com
* Correspondence: hanzhiyou@gmail.com (Z.H.); masa.inaba@nifty.com (M.I.)

Citation: Han, Z.; Kida, K.; Katsumata, K.S.; Handa, M.; Inaba, M. Effect of Hemicellulose on the Wet Tensile Strength of *Kozo* Paper. *Molecules* **2023**, *28*, 6996. https://doi.org/10.3390/molecules28196996

Academic Editors: Yi Tang, Yueer Yan and Yuliang Yang

Received: 25 August 2023
Revised: 30 September 2023
Accepted: 3 October 2023
Published: 9 October 2023

Copyright: © 2023 by the authors. Licensee MDPI, Basel, Switzerland. This article is an open access article distributed under the terms and conditions of the Creative Commons Attribution (CC BY) license (https:// creativecommons.org/licenses/by/ 4.0/).

Abstract: *Kozo* paper, *usu-mino-gami*, is frequently used as the first back lining paper of hanging scrolls in order to support the main paper with a painting or a work of calligraphy on it. To dye it an appropriate color, paper is often treated with an alkali mordant solution. However, current *kozo* paper products have received such comments from conservators that wet tensile strength is weak and hard to handle. Therefore, improving the wet tensile strength of *kozo* paper is required. In previous papers, the effect of the sheet forming method, cooking condition, and parenchyma cell content between fibers on the wet tensile strength of *kozo* paper has been investigated. In this paper, the effect of glucuronoxylan, the main component of hardwood hemicellulose on the wet tensile strength of *kozo* paper was investigated. The wet tensile strength of *kozo* paper, when made in different cooking conditions, was evaluated using the Finch device. Glucuronoxylan content in fiber was analyzed using GC-FID. According to the results, it has been proved that glucuronoxylan content (with a xylan to glucan molar ratio of 4.43% to 5.16%) itself contributes to the wet tensile strength of the *kozo* sheet. Therefore, to increase the wet tensile strength of *kozo* paper, it is recommended to cook under milder conditions, thus retaining a higher amount of glucuronoxylan in the pulp.

Keywords: paper-based cultural heritage; wet tensile strength; Japanese paper; xylan; hemicellulose; mounting paper; analysis techniques

1. Introduction

In Japanese mounted paintings, a paper-based backing is used to provide direct support for the delicate main paper. This backing sheet, called *"hada-ura-gami"*, is especially used with hanging scrolls, which commonly employ a thin grade of *"usu-mino-gami"* (Japanese *kozo* paper) as their primary lining. This first backing paper is often dyed to achieve a better tone and color of the image on the main paper and is adhered during mounting; therefore, a high wet tensile strength is essential.

Usu-mino-gami, made by the late Japanese papermaker Kozo Furuta, gained renown for its impressive wet strength. The paper made by his apprentice, Satoshi Hasegawa, was comparatively weak, leading Hasegawa to seek out the reason for this disparity. The resulting research lays the foundation for the present study.

In the initial study [1], we analyzed sheet samples that had been prepared under various cooking conditions by different sheet formers and produced using different sheet-forming techniques. The wet tensile strength of these samples was then assessed using the Finch method. The results show that wet tensile strength became higher with milder cooking conditions, such as a shorter cooking time and mild alkali cooking agent.

In the subsequent investigation [2], we explored the relationship between wet tensile strength and parenchyma cell content (defined as the proportion of the film area located between fibers on the surface area of the sheet). Our examination encompassed *kozo* sheets that had been prepared with identical alkali concentrations (of either sodium carbonate or sodium hydroxide) but different cooking times, as well as *kozo* sheets produced using the same cooking durations but with different alkali concentrations. Additionally, we evaluated *kozo* sheets made while maintaining the same cooking conditions but gradually reducing the parenchyma cell content using a fiber classier (screen opening 75 μm). Our findings revealed a direct correlation between higher wet tensile strength and increased parenchymal cell content.

The main chemical components of *kozo* paper are cellulose, hemicellulose, lignin, and pectin. According to Ishii [3], the xylem of hardwood typically contains approximately 20–30% hemicellulose, of which 80–90% is O-acetyl-4-O-methylglucuronoxylan (also known as glucuronoxylan). This is followed by glucomannan, the next most abundant hemicellulose found in hardwood. Since *kozo* paper originates from ligneous bark, it is likely that its composition is similar to the aforementioned hemicelluloses.

Some of the hemicellulose in wood becomes degraded and is lost during the cooking process. This occurs because of the peeling reaction, a process in which the 1→4-linked polysaccharides of cellulose and hemicellulose are successively removed from their reducing end groups during alkali treatment. As a result, both compounds undergo a reduction in their degree of polymerization. Because the original degree of polymerization for cellulose is relatively high, the effect of depolymerization is small, although their yield is reduced. In the case of glucomannan, however, the original degree of polymerization is around 70, so this peeling reaction contributes significantly to a reduction in the degree of polymerization [4]. Consequently, relatively significant amounts of glucomannan are lost during cooking due to its low molecular weight [5]. Glucuronoxylan, by contrast, exhibits a higher resistance to degradation at temperatures below 140 °C. The peeling reaction stops at the residue with 4-O-methyl glucuronic acid at the C-2 position and acetyl groups at the C-2 and C-3 positions [6]. Danielsson [7] reported that glucuronoxylan is reabsorbed on the fiber when the alkali concentration decreases during the late stages of cooking, as well as when the temperature decreases. Danielsson and Lindstom [5] reported that when glucuronoxylan from birch wood is adsorbed onto softwood pulp, the resulting increase in the higher degree of polymerization as well as more carboxyl group densities of glucuronoxylan leads to greater adsorption onto the fiber surface. Additionally, Mitikka-Eklund et al. [8] confirmed that glucuronoxylan was absorbed on the outer surface of the fiber.

Schonberg et al. [9] reported that the addition of glucuronoxylan and its adsorption onto the fibrous surface increased tensile strength. However, they noted that the glucuronoxylan retained within the fiber did not affect tensile strength or tear strength. Ban et al. [10] stated that the damage to cellulose and hemicellulose caused by cooking has a greater impact on paper quality than the relationship between the tensile strength of the sheet and the fiber's hemicellulose content.

Robinson reviewed the effect of hemicellulose on fiber bonding as follows. Since hemicellulose is flexible, it provides more hydrogen bonds per unit area than the cellulose surface. Furthermore, hemicellulose has a molecular chain with small molecular weight and few steric obstructions; therefore, the hydroxyl groups of hemicellulose can easily bind with hydroxyl groups on the surface of cellulose, which lacks mobility [11].

In particular, glucuronoxylan has a side chain containing a carboxyl glucuronic acid, which gives rise to a stronger hydrogen bond with hydroxyl groups such as cellulose. As mentioned above, the amount of glucuronoxylan content, the main component of hardwood hemicellulose, is known to contribute to paper strength.

Furthermore, in the papermaking process, because fiber itself is relatively rough and elastic, it must be beaten before becoming pulp. The fiber bundles are broken up to present fibrillation. Fibers also transfer to the appropriate length, and the surface also becomes gelatinized via a beating process. They expand and become plastic when exposed to water. Because hemicellulose has a strong hydrophilicity, when it is present on the surface of the fiber, water causes it to swell. The overall plasticity of the fiber is thus enhanced and becomes gelatinized more easily. When making paper, the strength of the paper increases due to the increase in the combination of strong plastic fibers [12].

In this paper, we began by examining glucuronoxylan content in the fibers of the *usu-mino-gami* made by Hasegawa. After this, we analyzed samples produced using different cooking methods in light of the findings on wet tensile strength gleaned from our earlier studies. We then investigated glucuronoxylan content following its extraction from *kozo* chips and its addition to *kozo* pulp when treated under identical cooking conditions and adsorbed onto the fibers. The purpose of this study was to clarify the effect of glucuronoxylan on the wet tensile strength of paper. It provides a guideline for the production of wet tensile strength *kozo* paper and contributes to the preservation and conservation of cultural properties with good paper quality.

2. Result and Discussion

2.1. Usu-Mino-Gami

Neutral carbohydrate components in Hasegawa-made *usu-mino-gami* were analyzed, with results normalized by molar ratio to glucan content (Table 1). Here, all the xylan was derived from glucuronoxylan, and the mannan was derived from glucomannan. Therefore, the neutral carbohydrate value indicates the relative amount of these hemicelluloses. Notably, the *usu-mino-gami* made by Furuta (Fu) exhibited not only a high glucuronoxylan content but also a high glucomannan content, indicating milder cooking conditions.

Table 1. Neutral carbohydrate composition of *usu-mino-gami* (molar ratio to glucan).

Molar Ratio to Glucan/%	Fu	W2(H)	S2(H)	N2(H)	N2(F)
Rhamnan	1.43 ± 0.01	1.34 ± 0.04	1.39 ± 0.02	1.05 ± 0.03	1.00 ± 0.01
Arabinan	0.54 ± 0.06	0.38 ± 0.02	0.43 ± 0.09	0.20 ± 0.04	0.24 ± 0.01
Xylan	4.87 ± 0.40	3.96 ± 0.01	3.74 ± 0.04	3.70 ± 0.08	3.70 ± 0.05
Mannan	1.76 ± 0.09	1.13 ± 0.10	1.08 ± 0.02	0.80 ± 0.02	0.61 ± 0.05
Galactan	1.87 ± 0.12	1.74 ± 0.03	2.15 ± 0.06	1.41 ± 0.01	1.57 ± 0.03
Yield of neutral carbohydrate */%	80.0 ± 0.6	82.7 ± 0.3	85.3 ± 0.4	83.8 ± 0.5	82.4 ± 0.6

* Including glucan yield.

Figure 1 shows the relationship between the wet tensile index (in 44 mmol/L potassium hydroxide solution) [1] and glucuronoxylan content (the molar ratio of xylan to glucan) of the *usu-mino-gami* reported in the first paper. As can be seen, a higher wet tensile index correlated with a higher glucuronoxylan content. Consistent with this finding, the sample with the highest wet tensile index, Fu, also had the highest glucuronoxylan content, while the sample with the lowest wet tensile index, N2(F), had the lowest glucuronoxylan content. These results suggest that weaker cooking conditions are conducive to enhancing the wet tensile strength of *usu-mino-gami* and that glucuronoxylan content increases in the process.

However, interpreting the results related to the Hasegawa sample proved challenging due to varying cooking conditions, specifically due to changes in the alkali concentrations and cooking time. Therefore, we had *kozo* chips from Kochi Prefecture and manufactured *kozo* sheets with the same alkali concentration or a constant cooking time. The experimental results for *kozo* sheets are shown below.

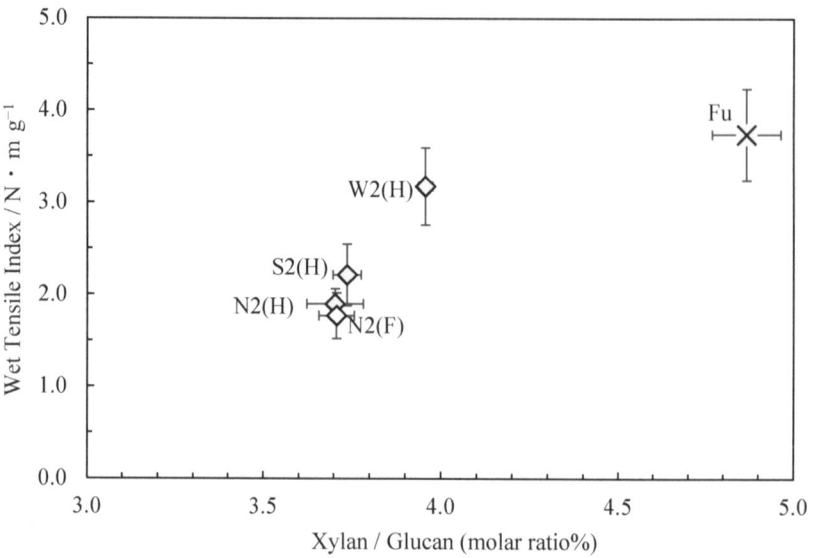

Figure 1. The relationship between xylan/glucan molar ratio and the wet tensile index for *usu-mino-gami*.

2.2. Kozo Sheets Produced with Identical Alkali Concentrations and Different Cooking Time

The results of the neutral carbohydrate analysis of *kozo* sheets prepared using identical alkali concentrations with different cooking durations are shown in Tables 2 and 3. The neutral carbohydrate content derived from all hemicelluloses was higher in the sample cooked in sodium carbonate (KC) than in the sample cooked in sodium hydroxide (KH) when the same cooking time was used. This finding aligns with previous research by Gustavsson, AL-Dajani [13], and Jiang et al. [14], who reported that in wood pulp, elevated alkalinity causes hemicellulose to degrade and dissolve more easily. Therefore, it comes as no surprise that the hemicellulose content in the fiber of samples cooked with sodium hydroxide (KH), which is highly alkaline, is notably diminished compared with that which was cooked with sodium carbonate (KC).

Table 2. Neutral carbohydrate composition of *kozo* sheets using sodium carbonate to cook *kozo* fiber.

Molar Ratio to Glucan/%	KC1	KC1.5	KC2	KC2.5	KC3.5
Rhamnan	1.18 ± 0.02	1.12 ± 0.01	1.14 ± 0.03	1.01 ± 0.02	1.02 ± 0.03
Arabinan	0.23 ± 0.06	0.23 ± 0.01	0.26 ± 0.02	0.22 ± 0.00	0.23 ± 0.00
Xylan	3.93 ± 0.01	3.93 ± 0.02	3.91 ± 0.05	4.15 ± 0.01	4.13 ± 0.08
Mannan	1.12 ± 0.03	1.04 ± 0.03	0.94 ± 0.02	0.95 ± 0.07	0.85 ± 0.02
Galactan	1.56 ± 0.03	1.54 ± 0.01	1.48 ± 0.00	1.40 ± 0.01	1.42 ± 0.01
Yield of neutral carbohydrate */%	81.5 ± 0.6	81.9 ± 0.1	81.5 ± 0.2	79.9 ± 0.2	82.8 ± 0.1

* Including glucan yield.

Table 3. Neutral carbohydrate composition of *kozo* sheets using caustic sodium hydroxide to cook *kozo* fiber.

Molar Ratio to Glucan/%	KH0.5	KH1	KH1.5	KH2	KH2.5
Rhamnan	0.80 ± 0.03	0.71 ± 0.01	0.71 ± 0.01	0.69 ± 0.00	0.70 ± 0.01
Arabinan	0.10 ± 0.01	0.10 ± 0.02	0.09 ± 0.01	0.08 ± 0.00	0.12 ± 0.04
Xylan	3.55 ± 0.08	3.49 ± 0.11	3.53 ± 0.02	3.34 ± 0.01	3.54 ± 0.15
Mannan	0.61 ± 0.01	0.43 ± 0.01	0.34 ± 0.02	0.34 ± 0.03	0.33 ± 0.04
Galactan	0.85 ± 0.01	0.65 ± 0.00	0.53 ± 0.00	0.48 ± 0.01	0.56 ± 0.02
Yield of neutral carbohydrate */%	80.2 ± 0.3	77.8 ± 0.5	81.0 ± 0.2	83.0 ± 0.2	78.6 ± 0.9

* Including glucan yield.

The relationship between the *kozo* sheets' glucuronoxylan content and pulp yield is shown in Figure 2. Although Gustavsson, AL-Dajani [13] and Jiang et al. [14] reported a decrease in glucuronoxylan content with longer cooking times, Clayton and Stone [15] stated that since glucuronoxylan is reabsorbed following cooking, cooking times have little impact on the final content. In this study, the glucuronoxylan content was almost the same with each alkali, regardless of the cooking time.

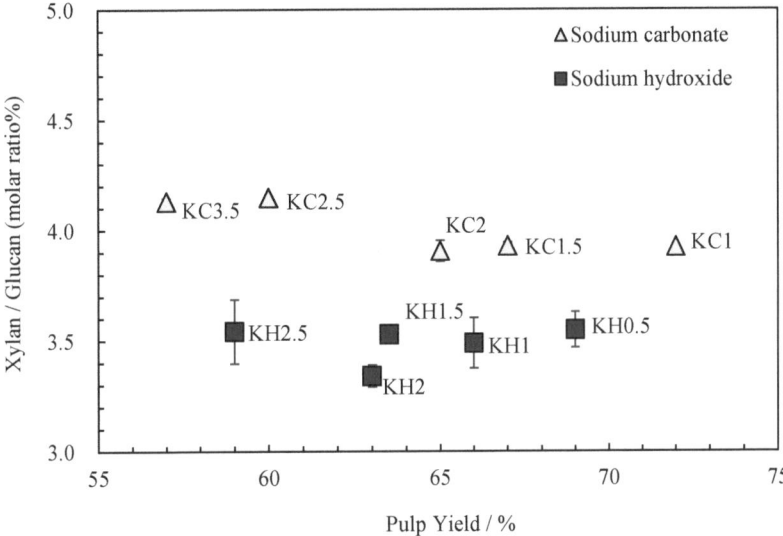

Figure 2. The relationship between xylan/glucan molar ratio and the pulp yield for *kozo* sheets prepared using different cooking agents and times.

Figure 3 shows the relationship between the glucuronoxylan content and wet tensile index. While the wet tensile index was found to decrease as cooking time increased, glucuronoxylan content changed little when the same cooking agent was employed. This finding underscores the role of other factors, such as parenchyma cell content and the degree of cellulose damage, in influencing changes in strength [2]. However, it should be noted that the amount of glucuronoxylan content varied when different cooking agents were used, directly impacting wet tensile strength.

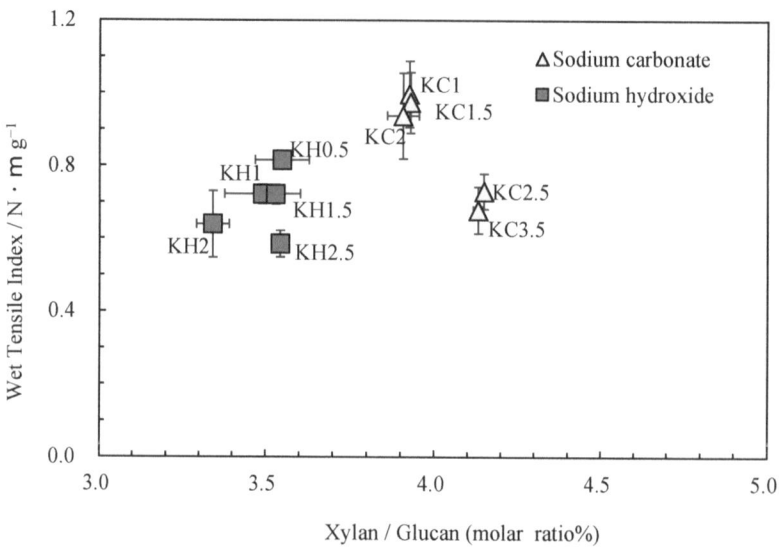

Figure 3. The relationship between xylan/glucan molar ratio and wet tensile index for *kozo* sheets.

2.3. Kozo Sheets Produced with Uniform Cooking Times but Different Alkali Concentration

Neutral carbohydrate analysis was performed on the *kozo* sheets by varying the sodium carbonate concentration while maintaining the same cooking time (Table 4). Figure 4 shows the relationship between the pulp yield and glucuronoxylan content. Among KC1-10 (KC1 in the previous chapter), KC1-12, KC1-14, and KC1-16, the glucuronoxylan content tended to decrease as the alkali concentration increased. Figure 5 shows the relationship between these data and the wet tensile index [2], where it can be observed that a higher wet tensile index is correlated with lower alkali concentrations. Additionally, wet tensile strength is linked to the amount of glucuronoxylan content.

Table 4. Neutral carbohydrate composition of *kozo* sheets prepared using different sodium carbonate concentrations to cook *kozo* fiber.

Molar Ratio to Glucan/%	KC1-10 [*1]	KC1-12	KC1-14	KC1-16
Rhamnan	1.18 ± 0.02	1.21 ± 0.08	1.19 ± 0.04	1.17 ± 0.06
Arabinan	0.23 ± 0.06	0.27 ± 0.01	0.18 ± 0.01	0.20 ± 0.01
Xylan	3.93 ± 0.01	3.95 ± 0.10	3.76 ± 0.10	3.81 ± 0.07
Mannan	1.12 ± 0.03	1.12 ± 0.05	1.05 ± 0.04	1.00 ± 0.10
Galactan	1.56 ± 0.03	1.59 ± 0.03	1.42 ± 0.02	1.42 ± 0.10
Yield of neutral carbohydrate [*2]/%	81.5 ± 0.6	92.3 ± 0.2	93.4 ± 0.2	92.9 ± 0.3

[*1] Sample KC1-10 is the same sample as KC1 in the previous chapter. [*2] Including glucan yield.

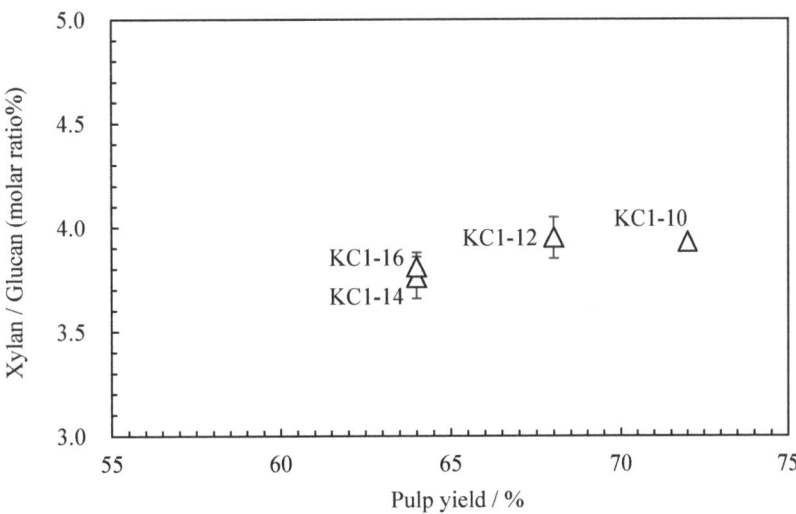

Figure 4. The relationship between pulp yield and xylan/glucan molar ratio of sodium carbonate when cooking *kozo*.

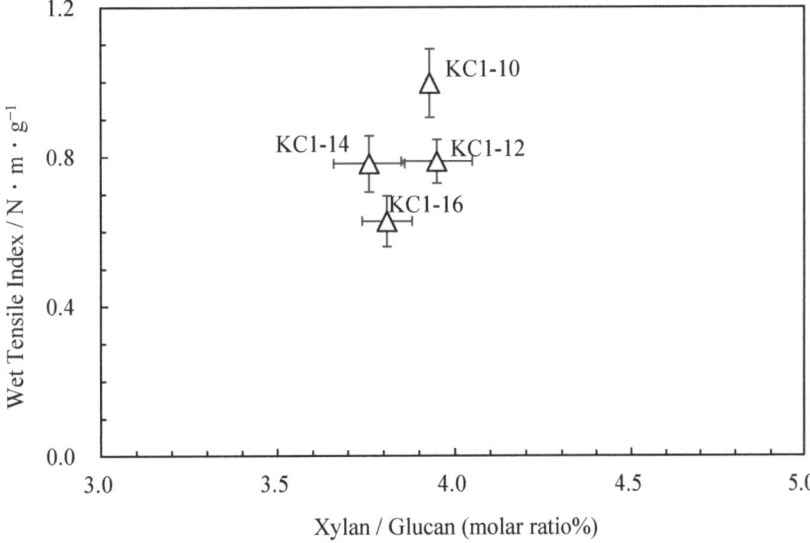

Figure 5. The relationship between xylan/glucan molar ratio and wet tensile index for *kozo* sheets of sodium carbonate when cooking.

2.4. Kozo Sheets Made with Absorbing Glucuronoxylan

The above experiments indicate a correlation between the glucuronoxylan content and the wet tensile index. However, different cooking conditions not only affect the content of hemicellulose but also cause varying degrees of damage to cellulose and parenchyma cell content, which probably also leads to an effect on paper strength. Therefore, in the following paragraph, in order to explore the relationship between glucuronoxylan content and wet tensile strength, cooking conditions were unified to ensure that glucuronoxylan content was the only variable.

Therefore, a strategy was employed in which glucuronoxylan extracted from *kozo* chips was added to *kozo* pulp that had been cooked under identical conditions. The glucuronoxylan content was then adsorbed onto the fibers, resulting in *kozo* paper samples that differed only in their glucuronoxylan content, allowing for its specific effect on paper strength to be investigated. The results of the neutral carbohydrate analysis are shown in Table 5. Though glucuronoxylan content was added just before cooking concluded, the proportion measured in the samples only ranged from 4.43% to a maximum of 5.16% for the xylan to glucan molar ratio. Figure 6 shows that the addition of glucuronoxylan content increased the wet tensile index, and although this phenomenon is known to occur in the production of wood pulp-based paper, our results confirm that a similar effect takes place when glucuronoxylan content is adsorbed onto the fibers of *kozo* paper.

Table 5. Neutral carbohydrate composition of *kozo* sheets made with absorbing glucuronoxylan.

Molar Ratio to Glucan/%	HC0	HC0.5	HC2.5
Rhamnan	1.10 ± 0.04	1.18 ± 0.01	1.12 ± 0.04
Arabinan	0.34 ± 0.02	0.31 ± 0.03	0.37 ± 0.03
Xylan	4.43 ± 0.15	4.69 ± 0.04	5.16 ± 0.04
Mannan	1.36 ± 0.06	1.33 ± 0.02	1.36 ± 0.08
Galactan	2.04 ± 0.06	2.07 ± 0.03	1.99 ± 0.01
Yield of neutral carbohydrate */%	85.0 ± 0.3	85.2 ± 0.3	86.7 ± 0.3

* Including glucan yield.

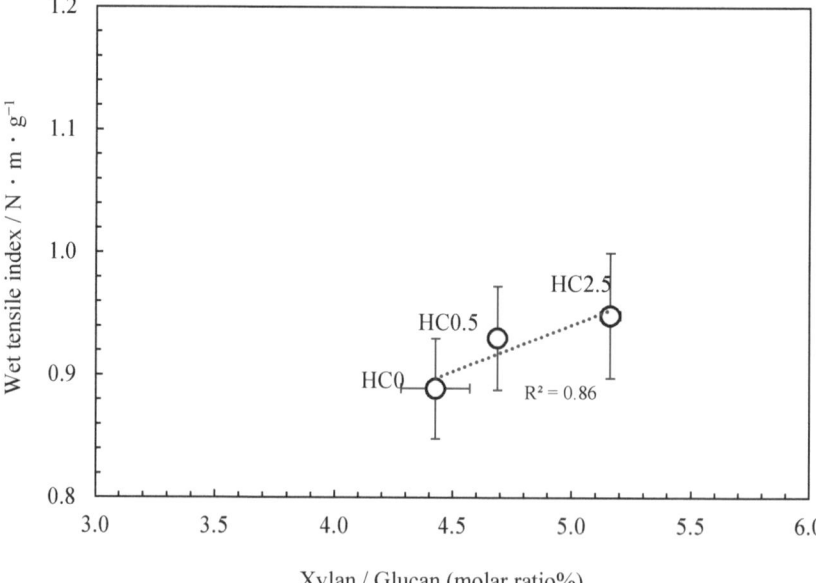

Figure 6. The relationship between xylan/glucan molar ratio and the wet tensile index for *kozo* sheets when xylan content is added.

3. Experimental Section

3.1. Sample

3.1.1. Hasegawa's *Usu-Mino-Gami*

Five samples were prepared at Hasegawa Washi Kobo, Satoshi Hasegawa's paper mill (Table 6). Nasu *kozo* white bark was cooked using sodium carbonate (Na_2CO_3) to form a sheet. Sample N2 represents a specimen made by Hasegawa according to his usual cooking process and conventional papermaking techniques. For sample N2(F), the same stock was bag-washed to remove parenchyma cells and fine fibers. The creation of sample S2 entailed an increase in cooking agent concentration and a shorter cooking time, while for sample W2, the cooking agent concentration was decreased, but the cooking time was likewise short. The final sample, Fu, was made by the late Furuta [1].

Table 6. Cooking conditions of *Usu-mino-gami* *.

Sample	Production Year	Cooking Chemical	Alkaline Concentration/%	Cooking time/h	Duration before Washing/h	Gramarge [*4] /g m^{-2}	Thickness [*4] /μm
S2 [*1]	2012	Na_2CO_3	10, 40	1(10%) + 1/6(40%)	12	15.9 ± 0.6	55.3 ± 4
N2 [*1]			13	2		14.8 ± 0.9	51.1 ± 4
N2(F) [*1]						17.7 ± 1.1	62.9 ± 4
W2 [*1]			10	1		16.7 ± 0.6	57.7 ± 3
Fu [*2]	Before 1994	Na_2CO_3	(12–13) [*3]	(1) [*3]	(2–4) [*3]	9.8 ± 0.8	48 ± 4

* This table is copied from a previous paper [1]. [*1] Samples were prepared by Mr. Satoshi Hasegawa. N2(F) was prepared from washed pulp (N2) in a cloth bag (*Fukuroarai*) with water before sheet preparation. [*2] Fu was prepared by Mr. Kozo Furuta, one of the best *usu-mino-gami* craftsmen ever. [*3] According to the survey by Yagihashi [16]. [*4] Mean ± standard deviation (N = 10).

3.1.2. Kozo Sheets Produced with Identical Alkali Concentrations or Uniform Cooking Time

For the preparation of *usu-mino-gami*, the alkali concentration and cooking time were adjusted to vary the cooking strength. As shown in Table 7, these *kozo* sheets were prepared according to the Tappi standard handmade method (JIS P8222:2005) [17] with a grammage of around 30 g/m^2 [2].

Table 7. Cooking conditions of *kozo* sheet *.

Sample	Cooking Chemical	Alkaline Concentration/%	Cooking Time/h	Duration before Washing/h	Gramarge [*1] /g m^{-2}	Thickness [*1] /μm
KC1 (KC1-10)	Na_2CO_3	10	1	1	26.8 ± 0.9	64 ± 3
KC1.5			1.5		29.7 ± 1.4	64 ± 6
KC2			2		29.5 ± 1.6	68 ± 5
KC2.5			2.5		34.8 ± 1.8	66 ± 4
KC3.5			3.5		33.2 ± 2.1	68 ± 4
KC1-12		12	1		28.0 ± 1.5	52 ± 2
KC1-14		14	1		27.1 ± 1.4	53 ± 1
KC1-16		16	1		26.6 ± 1.2	53 ± 2

Table 7. Cont.

Sample	Cooking Chemical	Alkaline Concentration/%	Cooking Time/h	Duration before Washing/h	Gramarge [*1] /g m^{-2}	Thickness [*1] /μm
KH0.5			0.5		26.5 ± 1.9	59 ± 5
KH1			1		25.8 ± 1.5	62 ± 3
KH1.5	NaOH	10	1.5		31.8 ± 2.4	65 ± 3
KH2			2		29.1 ± 2.0	58 ± 4
KH2.5			2.5		28.9 ± 2.6	60 ± 5

* This table is copied from a previous paper [2]. [*1] Mean ± standard deviation (N = 10).

3.1.3. Samples with Different Glucuronoxylan Content under Identical Cooking Conditions

Glucuronoxylan was isolated according to the following procedure. Kozo chips were extracted with methanol in a Soxhlet extractor for 6 h. The air-dried chips were then stirred in 4 L of distilled water for 2 days at room temperature to extract the water-soluble components. After additional air-drying, glucuronoxylan was extracted using 1 L of a 10% KOH solution for 3 h at room temperature under a nitrogen flow with agitation. Following suction filtration using ADVANCTEC No. 4A filter paper, the filtrate was introduced into 2.6 L of methanol containing 0.125 L acetic acid while stirring. The precipitated glucuronoxylan was collected via centrifugation and washed sequentially with 80% ethanol, anhydrous ethanol, and ethyl ether. The resulting crude glucuronoxylan was then re-dissolved in 1 L of a 5% KOH solution, purified via precipitation with 0.75 L ethanol, and the precipitate was again collected using centrifugation. It was then washed with ethanol and dried over phosphorus pentoxide under a vacuum in a desiccator using a rotary pump [18].

The following process was then used to add the glucuronoxylan to the pulp. First, pulp from Kochi *kozo* chips (60 g, cut into 4 mm lengths) was cooked in 1.8 L of 10% sodium carbonate solution for 50 min. Once it had cooled, the pulp was suction-filtered to separate the fiber from the cooking liquid. It was then divided into three equal parts according to weight. The cooking liquid was heated while nitrogen gas was blown into it. and when its boiling point was reached, glucuronoxylan was added, respectively, in amounts of 0.5 g and 2.5 g. These samples were named HC0.5 and HC2.5. The pulp was introduced and heated for 10 min, after which it was removed without cooling. The sample without the addition of glucuronoxylan underwent the same procedure (HC0). The pulp was washed as usual, and then it was beaten to 5000 revolutions in a PFI mill. *Kozo* sheets of gramarge at approximately 30 g/m² were formed, as shown in Table 8 (JIS P8222:2005) [2].

Table 8. Addition of glucuronoxylan to *kozo* pulp.

Sample	Cooking Chemical	Alkaline Concentration/%	Cooking Time/h	Amount of Glucuronoxylan Addition/g
HC0				0
HC0.5	Na$_2$CO$_3$	10	1	0.5
HC2.5				2.5

3.2. Measurement of Wet Tensile Strength by Finch Method

Consistent with our initial report [1], the wet tensile index was assessed using the Finch method, which involves immersion in a 44 mmol/L potassium carbonate solution for 20 s (JIS P8135:1998) [19]. In the Finch device, a sample piece turns around a horizontal rod, and its two ends are held by an upper clamp. A water vessel (container) can move up or down. By lifting up the container, a part of the sample piece around the horizontal rod is immersed in the solution and kept for a scheduled time. Tensile testing can be performed immediately after lowering the container, as in Figure 7. Ten pieces were measured for each sample, and the means and standard deviations of the measurement results have been used in this paper.

Wet tensile strength is expressed below:

$$S_w = 0.5 \times \frac{X}{W \times n}$$

S_w: Wet tensile strength (kN/m)
X: Maximum load until raptured (N)
W: Width of a sample piece (mm)
n: Number of layered sample pieces (n = 1 in our case)

Wet tensile strength when determined was divided on the basis of the weight of the sheet to obtain the wet tensile index.

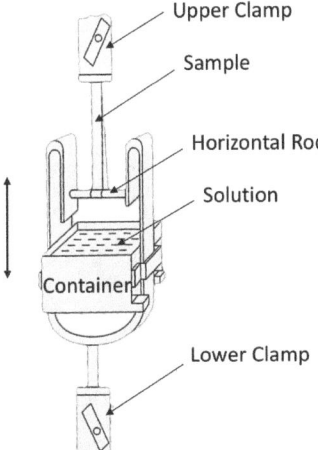

Figure 7. Finch device for short time wet strength measurement of paper (JIS P8135:1998).

3.3. Neutral Carbohydrate Analysis

Paper samples were disintegrated, and 50 mg of the oven-dried sample was weighed and subjected to sulfuric acid hydrolysis to form a monosaccharide. This was further reduced to alditol before being acetylated and analyzed using GC–FID [18]. The GC analysis conditions were as follows: Agilent 6890N; injection volume: 2 µL; column: Agilent J&W DB-225 (250 µm i.d., 30 m length, and 0.25 µm film thickness); oven temperature: 220 °C; analysis time: 30 min; inlet temperature: 250 °C; split ratio: 30:1; carrier gas: N_2, 0.7 mL/min; detector: FID (Flame Ionization Detector).

In this study, inositol was used as an internal standard; however, since acid hydrolysis is not sufficient, it is not expressed as an absolute amount but as a ratio of glucan. Two to three points for each sample were taken, and the entire process was performed. Measurement results are shown as the mean and standard deviation.

4. Conclusions

Although the amount of glucuronoxylan content in the *kozo* sheets varied with different alkali concentrations and types, its content remained quite consistent under conditions in which the type of alkali used and the alkali concentrations were identical.

Consequently, variables affecting the wet tensile index include the alkali used, the cooking intensity with sodium hydroxide or sodium carbonate, and the concentration of sodium carbonate assessed with consistent cooking times. All of these factors were found to impact the amount of glucuronoxylan content in *kozo* sheets, contributing to differences in their wet tensile strength.

With regard to the samples that had an addition of glucuronoxylan extracted from *kozo* chips to *kozo* pulp, it was confirmed that the wet tensile strength of the *kozo* sheet increased in proportion to the amount of glucuronoxylan that was added. Since the degree of cellulose damage and parenchyma cell content in these sheets was kept constant, it could finally be reliably determined that the glucuronoxylan content itself was the factor contributing to the wet tensile strength of *kozo* paper. Therefore, to increase *kozo* paper's wet tensile strength, sodium carbonate should be employed, and the fiber should be cooked under mild conditions, thus retaining a higher degree of glucuronoxylan content in the pulp. The cooking conditions that allow for the retention of a higher glucuronoxylan content are something to consider exploring in a future study.

The results of this paper intend to disclose information about the production of paper producers in the country. This allows conservators and producers to exchange information on issues such as defects in the use of the paper. We are convinced that the multifaceted discussions among producers, conservators, and researchers are based on objective data on the correlation between the raw materials and the manufacturing process of paper. The usability and stability of this paper provide significant support for the stable production of mounting paper.

Author Contributions: Z.H.: investigation, analysis, writing draft and editing; K.K.: analysis and review; K.S.K.: preparation of glucuronoxylan and review; M.H.: investigation and review; M.I.: project administration, investigation, review, and editing. All authors have read and agreed to the published version of the manuscript.

Funding: This research was partially funded by the JSPS Grant for Aid in Scientific Research (B) (23300323) "Development of Methods for Producing Durable Kozo Paper", the Manabu Yoshida Memorial Scientific Research Grant for Cultural Properties (2017-2)"Study to Improve Wet Strength of *Kozo* Paper Used for Conservation of Cultural Properties", and Conservation and Cultural Study Project-Subject Research Project-Belt and Road Joint Laboratory-China-Greece Demonstration Study on Joint Application of Cultural Heritage Conservation Technology (WB1070010).

Institutional Review Board Statement: Not applicable.

Informed Consent Statement: Not applicable.

Data Availability Statement: Not applicable.

Acknowledgments: We would like to express our sincere gratitude to Satoshi Hasegawa of Hasegawa Washi Kobo for his cooperation in the production of *usu-mino-gami* for this study. In addition, we would like to special thank Qu Liang of the Palace Museum for his support and assistance for this study. Finally, let us express our appreciation to the Mercedes-Benz Star Fund for its philanthropic financial support for this study.

Conflicts of Interest: The authors declare no conflict of interest.

Sample Availability: Not applicable.

References

1. Inaba, M.; Hasegawa, S.; Handa, M.; Enomae, T.; Taashima, A.; Han, Z.; Someya, S. Effect of Sheet Forming Method on Wet Tensile Strength of *Usu-mino-gami* (Japanese *Kozo* Paper). *Jpn. Tappi J.* **2019**, *73*, 559–574. (In Japanese + Full English Translation) [CrossRef]
2. Han, Z.; Kida, K.; Handa, M.; Inaba, M. Effect of Cooking Method on Wet Tensile Strength of Kozo Paper. *Jpn. Tappi J.* **2020**, *74*, 921–935. (In Japanese + Full English Translation)
3. Ishii, T. Hemiserurosu no Kagakukozo (The Chemical Structure of Hemicellulose). In *Mokushitsu no Kagaku (Wood Chemistry) Nihon Mokuzai Gakkai (The Japan Wood Research Society)* Ed.; Bun-eido Shuppan: Tokyo, Japan, 2010; pp. 123–126.
4. Watanabe, R. Hemiserurosu no Han-no to Riyou (Reaction and Application of Hemicellulose). In *Mokushitsu no Kagaku (Wood Chemistry) Nihon Mokuzai Gakkai (The Japan Wood Research Society)* Ed.; Bun-eido Shuppan: Tokyo, Japan, 2010; pp. 136–153.
5. Danielsson, S.; Lindstrom, M. Influence of Birch Xylan Adsorption during Kraft Cooking on Softwood Pulp Strength. *Nord. Pulp Pap. Res. J.* **2005**, *20*, 436–441. [CrossRef]

6. Sjöström, E. Koyojyu no Hemiserurosu (Hemicellulose in Hard Wood). In *Mokuzai Kagaku Kiso to Ouyou (Wood Chemistry Fundamentals and Applications)*; (Originally Published by Academic Press 1981); Kondo, T.; Samejima, K.; Higaki, M.; Kodansha, Translators; Academic Press: Cambridge, MA, USA, 1983; pp. 62–63.
7. Danielsson, S. *Xylan Reactions in Kraft Cooking and Their Influence on Paper Sheet Properties*; KTH Publish: Stockholm, Sweden, 2006; p. 53.
8. Mitikka, M.; Teeäär, R.; Tenkanen, M.; Laine, J.; Vuorinen, T. Sorption of xylans on cellulose fibers. In Proceedings of the 8th International Symposium on Wood and Pulping Chemistry, Helsinki, Finland, 6–9 June 1995; Volume 3, pp. 231–236.
9. Schonberg, C.; Oksanen, T.; Suurnakki, A.; Kettunen, H.; Buchert, J. The Importance of Xylan for the Strength Properties of Spruce Kraft Pulp Fibres. *Holzforschung* **2001**, *55*, 639–644. [CrossRef]
10. Ban, W.; Chen, X.; Andrews, G.; Heiningen, A. Influence of Hemicelluloses Pre-extraction and Re-adsorption on Pulp Physical Strength II. Beat Ability and Strength Study. *Cellul. Chem. Technol.* **2011**, *45*, 633–641.
11. Robinson, J.V. *Effect of Hemicellulose on Fiber Bonding, Pulp and Paper Chemistry and Chemical Technology*, 3rd ed.; Casey, J.P., Ed.; Wiley Interscience: New York, NY, USA, 1980; Volume II, pp. 940–944.
12. Machida, S. Seishi to Hemiserurosu ni Kansuru Ichikosatsu (The Study on Paper Making and Hemicellulose). *Jpn. Tappi J.* **1958**, *12*, 707–710.
13. Gustavsson, C.; Al-Dajani, W. The Influence of Cooking Conditions on the Degradation of Hexenuronic Acid, Xylan, Glucomannan and Cellulose during Kraft Pulping of Softwood. *Nord. Pulp Pap. Res. J.* **2000**, *15*, 160–167. [CrossRef]
14. Jiang, J.; Kettunin, A.; Henricson, K.; Hankaniemi, T.; Vuorinen, T. Effect of Alkali Profiles on Carbohydrate Chemistry during Kraft Pulping of Hardwoods. In Proceedings of the 10th International Symposium on Wood and Pulping Chemistry, Yokohama, Japan, 7–10 June 1999; Volume 1, pp. 406–411.
15. Clayton, W.; Stone, E. The Redisposition of Hemicelluloses during Pulping. Part I. The Use of a Tritium-labelled Xylan. *Pulp Pap. Mag. Can.* **1963**, *64*, 459–468.
16. Yagihashi, S. *Washi Fuudo Rekkishi Gihou (Japanese Paper Natural Environment, History and Technique)*; Kodansya Co.: Tokyo, Japan, 1981.
17. JIS P8222:2005; Pulps—Preparation of Laboratory Sheets for Physical Testing—Conventional Sheet-Former Method. Japanese Industrial Standard (JIS): Tokyo, Japan, 2005.
18. Shimizu, K. *GX (Gulukurono Kishiran) no Tanri (Isolation of Glucuronoxylan) in Shokubutsu Saibo Jikken Ho (Experimental Methods for Plant Cell Walls)*; Ishii, T., Ishizu, T., Umezawa, T., Kato, Y., Kishimoto, S., Konishi, Matsunaga, T., Eds.; Hirosaki Daigaku Syuppannkai: Aomori, Japan, 2017; p. 264.
19. JIS P 8135:1998; Testing Method of Wet Tensile Breaking Strength of Paper and Paperboard. Japanese Industrial Standard (JIS): Tokyo, Japan, 1998.

Disclaimer/Publisher's Note: The statements, opinions and data contained in all publications are solely those of the individual author(s) and contributor(s) and not of MDPI and/or the editor(s). MDPI and/or the editor(s) disclaim responsibility for any injury to people or property resulting from any ideas, methods, instructions or products referred to in the content.

Article

Research on the Structure and Properties of Traditional Handmade Bamboo Paper During the Aging Process

Zirui Zhu [1,2,†], Kai Zhang [1,†], Yu Xue [1], Zhongming Liu [1], Yujie Wang [3], Yanli Zhang [4], Peng Liu [1,2,*] and Xingxiang Ji [1,*]

[1] State Key Laboratory of Biobased Material and Green Papermaking, Qilu University of Technology, Shandong Academy of Sciences, Jinan 250353, China
[2] Institute for Preservation and Conservation of Chinese Ancient Books, Fudan University Library, Fudan University, Shanghai 200433, China
[3] Montverde Academy Shanghai, Shanghai 201318, China
[4] School of Chemistry and Chemical Engineering, Shanghai University of Engineering Science, Shanghai 201620, China
* Correspondence: liupengfdu@fudan.edu.cn (P.L.); xxjt78@163.com (X.J.)
† These authors contributed equally to this work.

Citation: Zhu, Z.; Zhang, K.; Xue, Y.; Liu, Z.; Wang, Y.; Zhang, Y.; Liu, P.; Ji, X. Research on the Structure and Properties of Traditional Handmade Bamboo Paper During the Aging Process. *Molecules* **2024**, *29*, 5741. https://doi.org/10.3390/molecules29235741

Academic Editors: Alejandro Rodríguez Pascual and Carmelo Corsaro

Received: 21 September 2024
Revised: 2 December 2024
Accepted: 3 December 2024
Published: 5 December 2024

Copyright: © 2024 by the authors. Licensee MDPI, Basel, Switzerland. This article is an open access article distributed under the terms and conditions of the Creative Commons Attribution (CC BY) license (https://creativecommons.org/licenses/by/4.0/).

Abstract: Handmade papers, as carriers of paper-based cultural relics, have played a crucial role in the development of human culture, knowledge, and civilization. Understanding the intricate relationship between the structural properties and degradation mechanisms of handmade papers is essential for the conservation of historical documents. In this work, an artificial dry-heat-accelerated aging method was used to investigate the interplay among the mechanical properties of paper, the degree of polymerization (DP) of cellulose, the chemical composition, the hydrogen bond strength, the crystallinity, and the degree of hornification for paper fibers. The results demonstrated for the first time that the mechanical properties of handmade bamboo paper exhibited an initial plateau region, a rapid decline region, and sometimes a second plateau region as it undergoes a dry-heat aging process. The changes in cellulose, hemicellulose, and lignin content were tracked throughout these three stages. The lignin content was relatively stable, while the cellulose and hemicellulose content decreased, which was consistent with the observed decline in mechanical properties. When the DP of cellulose decreased to the range of 600–400, there was a critical point in the mechanical properties of the paper, marking a transition from the initial stable region to a rapid decline region. The fiber embrittlement caused by cellulose chain breakage resulting from the decrease in DP was counteracted by the enhancement of intermolecular hydrogen bonds and the hornification process. A second stable region appeared when the DP was less than 400, marking a transition from a balanced or slightly decreasing trend in the initial plateau region to a sharp decline. This study also discussed for the first time that the formation of the second plateau region may be due to the presence of hemicellulose and lignin, which hinder the further aggregation of cellulose and maintain the structural stability of the fiber cell. The findings of this study can provide guidance for improving ancient book preservation strategies. On the one hand, understanding how these components affect the durability of paper can help us better predict and slow down the aging of ancient books. On the other hand, specific chemical treatment methods can be designed to stabilize these components and reduce their degradation rate under adverse environmental conditions.

Keywords: handmade paper; critical DP; H-bond; hornification

1. Introduction

The saying "Paper lasts for a thousand years, while silk endures for eight hundred" highlights the significance of paper as a crucial medium for the transmission and development of human civilization [1]. Traditional Chinese paper, with a history spanning over 2000 years, has played an irreplaceable role in ancient book printing, calligraphy, painting arts, and other cultural relics [2]. The general manufacturing processes for traditional Chinese paper involve steeping, fermenting, washing, steaming, boiling, natural bleaching, pulping, sheet forming, pressing, and drying, employing mild treatment conditions to minimize adverse effects on plant fibers [2,3]. Papers manufactured throughout the long history of Chinese papermaking can be categorized into bast paper, bamboo paper, straw paper, and mixed-fiber paper (i.e., *Xuan* paper), and each is endowed with distinct characteristics [4]. The methods of making various types of paper differ due to the composition and properties of the raw materials, although the basic principles of pulping and papermaking are the same. Longer bast fibers give bast paper better strength, while the mixed paper is a blend of long bast fibers and short rice straw fibers, resulting in excellent effects for calligraphy and painting. Due to its high content of lignin and hemicellulose, bamboo raw material is used to produce a unique type of paper known as bamboo paper. Handmade bamboo paper holds great importance in traditional papermaking in China, with a rich history and a wide variety of categories. The craftsmanship of bamboo paper flourished during the Tang and Song dynasties, particularly in the Song dynasty, where it gained dominance due to its cost-effectiveness, favorable texture, and availability. Furthermore, bamboo paper's desirable properties, such as its flexibility and water absorption, made it popular in calligraphy and printing. The manual production of bamboo paper reached its peak during the Ming and Qing periods when it was utilized not only for daily writing but also extensively in the restoration and printing of ancient books, as well as in calligraphy and mounting. In 2006, the bamboo-paper-making process was recognized in the first batch of national intangible cultural heritage in China [5]. A survey indicates that some regions in Fujian, Jiangxi, and Zhejiang provinces still produce bamboo paper suitable for archival and ancient book restoration. However, there are also problems, such as the lack of emphasis on bamboo paper, a decline in its quality, a disconnect between production, supply, and marketing, and a lack of successors [6]. More importantly, the relatively short fibers and lower cellulose content of bamboo paper make it less durable than papers made from other materials, such as bast, making it more susceptible to degradation and necessitating specialized conservation and restoration efforts in the restoration and printing of ancient books [1]. Therefore, studying bamboo paper's aging behavior and preservation methods is crucial for prolonging the lifespan of ancient books and manuscripts and gaining insights into their conservation and restoration needs.

During the long-term preservation process, paper can degrade due to a combination of internal and external factors. Internal factors include acidic degradation products and excessive alkali reserves within the paper itself, while external factors consist of light, temperature, humidity, air pollutants in the environment, and the presence of inks, pigments, fillers, insects, and microorganisms on the paper [7–10]. To study the degradation of paper, accelerated aging experiments are always conducted in laboratories due to the slow degradation rate under natural conditions [11]. These experiments involve severe conditions, such as elevated temperatures and humidity levels, intense ultraviolet and visible radiation, and significant pollutant concentrations, which expedite the deterioration of paper. Additionally, environmental factors such as high levels of air pollution and adverse meteorological conditions have been shown to exacerbate the degradation process of materials, including paper [12]. In order to gain a deeper understanding of the mechanism behind paper deterioration, researchers often use pure cotton/cotton linter cellulose paper and bleached sulfite softwood/hardwood cellulose paper as model papers [13–15]. These model papers are composed predominantly of cellulose, with minimal hemicellulose and lignin content; however, traditional bamboo papers actually contain relatively higher amounts of hemicellulose and lignin, which may exhibit certain differences in aging behavior due to their distinct structural and functional properties.

Researchers have conducted studies on bamboo paper from various perspectives. Compared with the center of the paper pages, the edges of traditional Chinese bamboo paper pages undergo chemical changes through oxidation and photo-aging effects [16,17]. Aging experiments conducted for 72 h at 105 °C in nitrogen, air, and sealed preservation environments reveal that nitrogen storage exhibits the best anti-aging properties, followed by air storage, while sealed preservation performs the worst. Sealed storage is not ideal as it inhibits the release of the paper's volatile substances. Therefore, it is recommended to use storage equipment that is breathable to allow for air circulation while protecting documents [18].

In their research on bamboo paper aging, Chen and Ding found that handmade bamboo paper with minimal processing is more susceptible to yellowing, while excessive treatment can harm the fibers and impact the thermal stability of the paper [19]. They also developed a quantitative model based on changes in characteristic temperatures of pyrolysis to better evaluate the degree of bamboo paper aging [20]. Additionally, the pyrolysis characteristics of bamboo paper under various dry-heat-aging conditions were studied using thermogravimetric analysis, revealing a deterioration in the thermal stability of bamboo paper. The difference in characteristic temperatures of pyrolysis of bamboo paper, $\Delta T0.5$, was proposed as a parameter to evaluate the degree of bamboo paper aging, with an exponential relationship being established between $\Delta T0.5$ and the retention rate of the tensile index, leading to the development of a quantitative model for assessing bamboo paper aging [21].

A comparative analysis of uncooked and cooked bamboo paper focused on their dimensional stability- and durability-related physicochemical indicators [22]. This study revealed that while uncooked bamboo paper exhibited better dimensional stability, cooked bamboo paper demonstrated superior durability, making it more suitable for meeting the quality requirements of paper used in the restoration of ancient books. Samples of uncooked and cooked bamboo papers were obtained from three paper workshops located in the *Jiangle, Liancheng*, and *Changting* regions of Fujian province. The results indicated that although uncooked paper displayed improved dimensional stability, its durability was inferior to that of cooked paper, thus rendering uncooked paper a more suitable material for the restoration of ancient books.

Recently, scholars compared six types of traditional handmade Chinese paper, including bast paper, bamboo paper, and grass paper, and they analyzed their wet-heat-accelerated aging behavior and degradation mechanisms [1]. This study found that the type of raw fiber material directly affects the durability of handmade paper. At the molecular level, most degradation occurs in the non-crystalline regions of cellulose, leading to the breakage of glycosidic bonds, the generation of oxidative groups, and changes in hydrogen bond arrangements in cellulose. Bark paper, due to its specific molecular and supramolecular structure of cellulose, has a longer lifespan and better durability than bamboo paper and grass paper. At the supramolecular level, the cellulose in bast paper has higher crystallinity, providing better stability during the aging process. These results provide valuable information for understanding the degradation mechanisms of handmade paper from different fiber materials and provide a scientific basis for improving the durability of handmade paper. In addition, researchers have conducted accelerated aging experiments under different conditions (humidity or dryness, air or nitrogen) and different durations (2, 4, 10, 25, and 50 weeks) to simulate the chemical degradation of paper [23]. Further research has been conducted on the relationship between chemical degradation and mechanical degradation during the aging process of paper, especially the impact of fibers and fiber-to-fiber bonds on the embrittlement of paper. This emphasizes that the embrittlement of fibers themselves has a greater impact on the overall embrittlement of paper than the degradation of fiber-to-fiber bonds, and the critical threshold of cellulose's degree of polymerization (DP), below which the embrittlement of paper will significantly accelerate, was pointed out. These findings are of great significance for understanding and predicting issues with the durability and preservation of paper. However, they focus on pure cellulose as the research object, while ancient paper contains a certain amount of hemicellulose and lignin, which have a profound influence on the aging behavior of paper and should also be given attention.

Despite the significance of Chinese handmade papers, the understanding of the relationship between their structure and properties during the aging and degradation process is still limited. The variety of raw materials used in the Chinese papermaking industry, as well as the intricate craftsmanship employed, makes this issue even more complicated. Handmade paper made in traditional crafts often contains a certain amount of lignin and hemicellulose, which, in addition to cellulose, may affect the degradation behavior of the paper. The influence of fiber raw materials on the aging behavior of paper under wet-heat-accelerated aging conditions was investigated. It was found that the longer fiber lengths result in a larger length-to-diameter ratio and a reduced presence of fine fibers in the paper. This, in turn, leads to a higher DP and crystallinity in the original cellulose. As a result, the paper has a longer lifespan and improved durability [1]. A recent study showed that environmental factors may exhibit similar trends in their impacts on the aging process. For pure cellulose paper, mechanical performance gradually decreases above the critical polymerization degree (DPc), while below the DPc, mechanical performance deteriorates significantly [23]. However, traditional bamboo paper often contains higher lignin and hemicellulose content, and the impact of these two components on the durability of the paper has been a subject of ongoing debate. This study focuses on traditional bamboo paper derived from bitter bamboo, aiming to investigate the aging behavior of paper with high lignin and hemicellulose content. It was found that the mechanical properties of bamboo paper with high lignin and hemicellulose content exhibit three stages during accelerated aging in dry heat as the DP decreases: a first plateau region, a rapid decline region, and a second plateau region. This study, for the first time, tracked the changes in cellulose, hemicellulose, and lignin content throughout these three stages. It also explored the reasons for the formation of the second plateau region, which may be attributed to the presence of lignin and hemicellulose inhibiting further cellulose aggregation, thus maintaining structural stability. The findings of this research can enhance our comprehension of the relationship between the structure and properties during the aging and degradation process of ancient paper-based books, provide evaluations and guidance for the development of conservation agents for ancient books, and establish a scientific foundation for the production of more durable and long-lasting handmade paper.

2. Results and Discussion

2.1. Mechanical Properties

Research on the aging and degradation of paper is typically approached from two perspectives. The first perspective involves studying changes in paper properties over time [24,25]. While this approach provides a direct reflection of the changes in mechanical properties, it can be challenging to determine the underlying causes due to the influence of fiber bonding strength and the inherent strength of the fibers themselves. Factors such as the bonding strength and inherent strength of fibers are influenced by a multitude of factors, with the DP of cellulose playing a significant role in the mechanical properties of paper [26]. Other studies have directly examined the relationship between the reduction in cellulose's DP and the loss of paper's mechanical properties under various conditions, including temperature, humidity, and acidity [14,27,28]. The loss of mechanical properties serves as an indicator of paper degradation; however, it is not always a linear function of cellulose degradation. As the DP of cellulose decreases with aging, the breaking of molecular chains leads to fiber brittleness, whether through oxidation or hydrolytic degradation mechanisms [23]. Additionally, a DPc for cellulose (DPc~750, Mn) has been identified, beyond which the mechanical properties decrease significantly, irrespective of the type of mechanical testing conducted [24]. This study builds upon new insights into the relationship between the cellulose DP and the mechanical properties of paper.

Compared with machine-made paper, handmade paper exhibits less variation in fiber orientation during the papermaking process. However, it is possible to control the water flow and reduce fiber orientation through the vibration of the paper screen in manual papermaking. Nevertheless, the water flow still influences the orientation of fibers in the longitudinal and transverse directions, corresponding to the wire and bamboo patterns of the paper mold [2]. This indicates that handmade paper also displays anisotropy in the transverse and longitudinal directions. In this study, the mechanical properties in different directions are referred to as the LD (longitudinal direction) and TD (transverse direction).

Figure 1 depicts the relationship between the tensile, tear, and folding endurance properties of handmade bamboo paper and its DP. The results reveal several key characteristics. Firstly, unaged handmade bamboo paper is primarily oriented in the longitudinal direction, enabling it to bear and transmit more load in this direction. As a result, it exhibits a higher tensile index and folding endurance than the transverse direction. Due to the ease with which fibers can be pulled out along the paper's longitudinal direction, while those perpendicular to it hinder this process, the transverse tear index is greater than the longitudinal tear index.

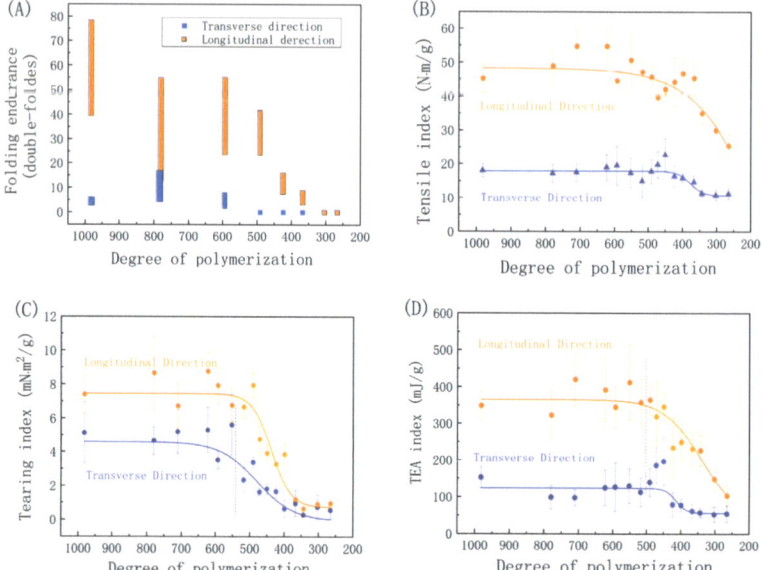

Figure 1. Evolution of mechanical properties (**A**) Folding endurance, (**B**) Tensile index, (**C**) Tearing index and (**D**) TEA index with the decrease in the degree of polymerization (DP) of cellulose, as measured using folding, tearing, and tensile tests on bamboo paper in the longitudinal and transverse directions under an accelerated dry-heating treatment at 105 °C. LD: longitudinal direction; TD: transverse direction. The vertical dotted line indicates the value of the critical DP.

The second characteristic is that the strength properties of the paper exhibit different patterns of decline with a reduction in cellulose DP caused by the extension of the dry-heat-aging time. In the early stages of dry-heat-accelerated aging (Figure 1A–D), the mechanical properties of the paper in the longitudinal direction nearly plateau for a period. However, in the middle and later stages of aging, the tensile index-TD (Figure 1B) and tear index-TD experience a significant decline, followed by a second plateau period (Figure 1C). On the other hand, the tensile index-LD gradually decreases without a second plateau period (Figure 1B). The trend of changes in the energy absorption index is consistent with the tensile index in both the longitudinal and transverse directions (Figure 1D).

Similarly to the phenomena reported in a previous study [23], this work also observed a DPc for the paper. Initially, when the paper's DP is relatively high, there is a plateau period in its properties. At this stage, cellulose degradation does not necessarily result in a loss of mechanical properties of the paper [29,30]. However, the mechanical properties of the paper decline after surpassing this DPc. The DPc for the paper's tensile properties and energy absorption index is estimated to be between 450 and 500 (Figure 1B,D), while for tearing performance, it is approximately between 500 and 600 (Figure 1C). Therefore, it can be inferred that the DPc for the plateau and decline periods in this study falls within the range of 450 to 600. This differs slightly from the previously reported Mn-750 [23], which is possibly due to the use of DP obtained from the viscosity-average molecular weight in this study, as well as variations in the raw material content for handmade paper, leading to different degradation processes and, consequently, affecting the DPc.

One consequence of cellulose polymer chain degradation, as the primary component of paper fibers, is an increase in paper brittleness during aging. This property could be assessed through folding endurance [27]. The folding endurance in the longitudinal direction (folding endurance-LD) exhibits a slight decline in the early stages of aging, with the DPc appearing at 500, after which a significant decline is observed (Figure 1A), consistently with previous findings [24]. On the other hand, the change in folding endurance in the transverse direction (folding endurance-TD) is not significant, remaining below 10 overall and dropping to 0 once the DP falls below 600. The folding endurance-TD is lower than 10 due to the paper's thinness and weaker transverse binding force, coupled with the decrease in fiber strength. In the longitudinal direction, where fibers have certain bonds, the folding endurance-LD gradually decreases along with the weakening of fiber strength. It sharply declines once the DP exceeds 500 and completely disappears when the DP falls below 350. The decline in folding endurance is mainly influenced by fiber brittleness caused by oxidation and cross-linking in the initial stage of aging [26].

2.2. DP and Component Contents

The DP is a crucial factor in evaluating the performance and longevity of paper [1,31,32]. Accelerated dry-heat aging can significantly decrease the DP of cellulose in paper, a process that is typically divided into three stages (Figure 2A). The initial stage occurs within 30 days, during which the DP of the paper decreases by nearly 40%. Subsequently, as the aging progresses to 50 days, the DP further drops by 10% to around 500. Although the DP continued to decline in this phase, the rate slowed down significantly. The sharp decline in DP results in extensive breakage of cellulose molecular chains, a decrease in fiber strength, and a notable reduction in folding endurance. However, it may not impact the bonding strength between fibers, thus maintaining tensile and tear strength, corresponding to the first plateau region. From 50 days onwards, as the paper continues aging up to 200 days, the DP decreases at a slower rate. Research has indicated that cotton fibers take approximately 850 days for the DP to decrease from 300 to 280, suggesting a slight decrease in molecular weight in the later stages of cellulose degradation [24]. Throughout the aging process from 50 to 200 days, there is a significant decline in the mechanical properties of the paper but a slow degradation of cellulose.

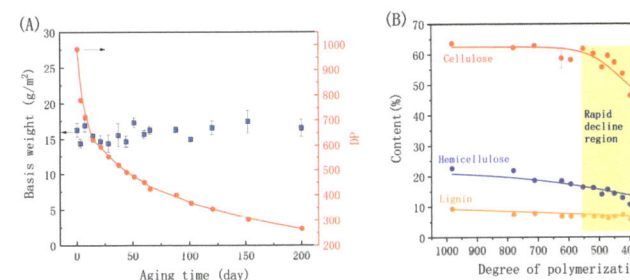
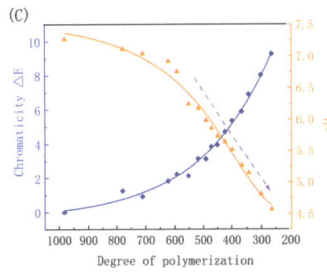

Figure 2. Variations in basic weight and DP as a function of aging time for bamboo paper (**A**); variations in chemical composition (**B**) and chromaticity ΔE and pH (**C**) with the decrease in DP under an accelerated dry-heating treatment at 105 °C.

It is often overlooked that during the aging process, the composition of the main components of paper, namely, cellulose, hemicellulose, and lignin, undergo constant changes.

As depicted in Figure 2B, the lignin content shows a gradual slight downward trend, particularly in the later stages of aging, where the content remains relatively stable when the DP is below 600. It is important to note that the cellulose and hemicellulose content exhibits a similar trend to that of the mechanical properties when comparing Figure 1 with Figure 2. In Figure 2B, a clear plateau region is evident for the cellulose and hemicellulose content before the DP exceeds 600, with a slight downward trend. Between DP of 600 and 400, a significant decrease is observed. Specifically, the cellulose content decreases from 60% to 50%, and the hemicellulose content drops from 18% to 12%. Once the DP falls below 400, the cellulose decrease rate slows down, while hemicellulose shows a plateau. Additionally, the paper's pH value gradually declines from an initial 7.25 to 6.8 when the DP is above 600. Subsequently, as the cellulose DP and content decrease, the pH value exhibits a linear downward trend (Figure 2C). Figure 2B illustrates that cellulose and hemicellulose are the main components undergoing degradation, leading to an increase in acidic degradation products. This acceleration of the acidification may be caused by the action of volatile organic compounds on cellulose or hemicellulose self-oxidation. During natural aging, paper undergoes color changes and becomes brittle, which is primarily due to the degradation of cellulose, the primary component of paper fibers [26]. Although the lignin content does not decrease significantly, it may play a role in these processes through changes in its functional groups. Understanding how the slowly changing cellulose, hemicellulose, and lignin content maintains mechanical stability when the DP exceeds 600, i.e., the first plateau region, warrants further investigation.

However, there is limited research explaining the DPc for cellulose [23]. This polymerization threshold is observed in traditional semi-crystalline polymers [33], ensuring a minimum amorphous phase thickness that influences the plastic deformation mechanism. The complexity of cellulose's microstructure adds to this issue [23,33]. The appearance of this first plateau period seems uncommon in aging studies of pure cellulose samples but emerges when the DP reaches a certain value [29].

2.3. Microstructures

The primary constituent of Chinese handmade paper is cellulose, which is a polymer consisting of linear chains of hundreds to thousands of D-glucose units linked via β-(1,4)-glycosidic bonds [34,35]. The hydroxyl groups present in cellulose participate in numerous intra- and intermolecular hydrogen bonds, resulting in various ordered crystalline arrangements [36]. These hydrogen bonds play a crucial role in the mechanical properties of paper by forming inter-fiber connections through interactions between the hydroxyl (–OH) groups in cellulose molecules [37]. The strength and quantity of hydrogen bonds can impact the interlayer spacing and elastic modulus of paper. Moreover, the presence of hydrogen bonds provides the paper with a self-healing capability, as these bonds can dynamically reform under certain conditions, repairing the microstructure of the paper. Some studies suggest that cellulose degradation at the supramolecular level leads to changes in the intensity of hydrogen bonds and the crystallinity of cellulose macromolecules [38,39].

Figure 3 depicts the alterations in hydrogen bond lengths and energies between and within cellulose molecules during the paper-aging process. As shown in Figure 3A,B, both intramolecular and intermolecular hydrogen bond energies increase throughout the aging process, with intermolecular hydrogen bonds displaying a more pronounced change. In the initial 40 days of aging, the bond length of intermolecular hydrogen bonds gradually decreases, indicating a potential movement of cellulose molecular chains toward each other. Subsequently, the bond length increases and stabilizes, fluctuating within a narrow range until day 200. The decrease in intramolecular hydrogen bond length and the increase in bond energy suggest possible contraction or distortion of molecular chains of cellulose due to the prolonged dry-heat treatment. Conversely, the intermolecular hydrogen bonds exhibit a significant decline followed by recovery, suggesting a potential rearrangement process of hydrogen bonds [40] that accompanies the decrease in the paper's cellulose DP. In the first plateau region of the paper's mechanical properties, despite the sharp decrease in DP, the increased hydrogen bond energy may strengthen the bonding between fibers, thereby mitigating the decay of the mechanical properties. In other words, while the cellulose molecular chains may experience breakage and make the fibers brittle, the enhanced strength of hydrogen bonds within the internal fiber network of the paper reinforces the binding between fibers and may lead to irreversible hornification.

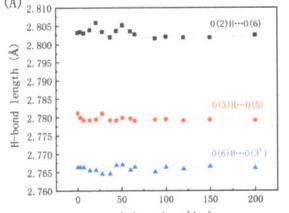 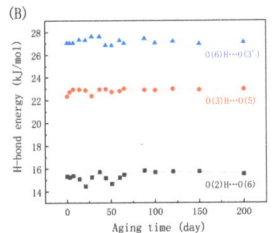 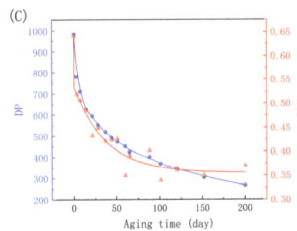

Figure 3. Variations in H-bond length (**A**), H-bond energy (**B**), DP, and water retention value (**C**) as a function of aging time for bamboo paper.

Hornification refers to the irreversible changes that occur in paper during the water removal process at either room temperature or high temperatures. These changes result in alterations in the paper's water absorption behavior, including reduced flexibility, decreased water retention capacity, and increased brittleness [41,42]. This phenomenon is attributed to the formation of irreversible hydrogen bonds between microfibrils within the fibers, which are typically associated with the durability and stability of paper. As hornification occurs, the flexibility of the paper decreases. This is because irreversible hydrogen bonds form between the microfibrils within the fibers during the hornification process, which restricts the relative movement of the fibers and reduces the flexibility of the paper. This is due to the breaking of cellulose molecular chains and a decrease in the DP during the aging process, resulting in weakened fiber-to-fiber bonding and increased fragility of the paper [42]. Additionally, hornification leads to a decrease in the water retention value (WRV) of the paper. When preserving and restoring paper-based cultural artifacts, such as ancient books and archives, the WRV is a crucial factor to consider. Papers with a higher WRV are usually more flexible because the moisture between the fibers reduces friction, making the paper easier to bend and fold without tearing. On the other hand, papers with a lower WRV are more prone to brittleness, as the lack of moisture weakens the fiber-to-fiber bonding, making the paper more susceptible to breakage when subjected to external forces. The dimensional stability of paper is also closely related to the WRV, as papers with a high WRV expand when absorbing moisture and contract when drying. A low WRV can lead to brittleness and easy damage to paper, which is crucial for long-term preservation and restoration strategies of cultural artifacts. It is generally believed that the WRV tends to decrease as paper ages, reflecting the degradation and structural changes in cellulose fibers. The findings presented in Figure 3C demonstrate that, similarly to the DP, the change in the WRV can be divided into three stages. However, when DP > 600, there is a clear overall correlation with the WRV, which sharply decreases as the DP decreases. After 50 days of aging, the DP experiences a slow decline, while the WRV exhibits a relatively stable fluctuation trend. Thus, it can be inferred that hornification displays an initial increasing trend followed by a leveling off throughout the entire aging process.

During dry-heat aging, the fibers gradually shrink from their initial cylindrical shape, with significant collapse and surface wrinkles being observed by day 28 (Figure 4a,b). Subsequently, the shrinkage slows down, but the overall fiber morphology remains stable (Figure 4c–f). Most fiber shrinkage is likely to occur within the first few hours of the drying process [43], which aligns with the trend reflected by the WRV. As the fiber structure contracts during the early stages of aging, the water absorption decreases. Once a certain level of contraction is reached, the fiber morphology remains almost unchanged, while the water absorption performance undergoes irreversible hornification.

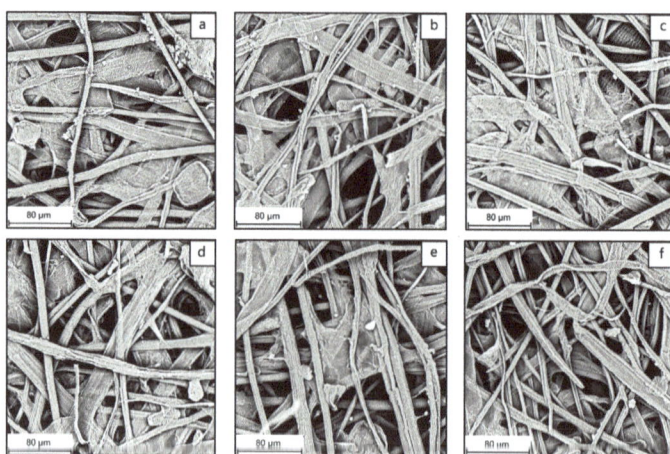

Figure 4. Morphological changes in bamboo paper during accelerated aging: (**a**) 0D, (**b**) 28D, (**c**) 50D, (**d**) 100D, (**e**) 150D, and (**f**) 200D (1000 times).

Research has also indicated a correlation between mechanical properties and crystalline structure during paper aging [16,44]. The supramolecular structure of cellulose affects the degradation of its molecular structure, with a higher supramolecular order hindering degradation [39,45]. Although aging causes a reduction in the adsorption and swelling capacity of the paper, leading to an increased crystallinity [41], the infrared OKI crystallinity index shown in Figure 5A demonstrates a significant decrease in the first 3 days, followed by an increase period (3–40 days), and then a decrease after 100 days, eventually leveling off within a certain range. The X-ray diffraction (XRD) results show an initially stable period, followed by regular fluctuations after 50 days. The main interplanar distances, especially for the 002 plane, remain generally stable. However, the crystal grain size gradually decreases from 8 layers to 5–7 layers in the later stages (Figure 5B). This change is likely attributed to the hornification effect caused by cellulose microfibrils during the drying process. In the later stages of paper aging, although thermal aging leads to fiber contraction or collapse, the amorphous regions of cellulose may have degraded significantly. Cellulose co-crystallization, which involves hydrogen bonds, hydrophobic interactions, and van der Waals forces within fibers, was not observed. It may be due to the inhibition of lignin and hemicellulose from the hornification of paper cellulose, preventing the enlargement of crystalline particles and the aggregation of cellulose microfibrils [46]. Moreover, hemicellulose and lignin may contribute to bonding and maintaining the stability of the fiber structure despite the decrease in the cellulose DP. Xylan and glucomannan are important types of hemicellulose that have been found to reduce hornification [43,47].

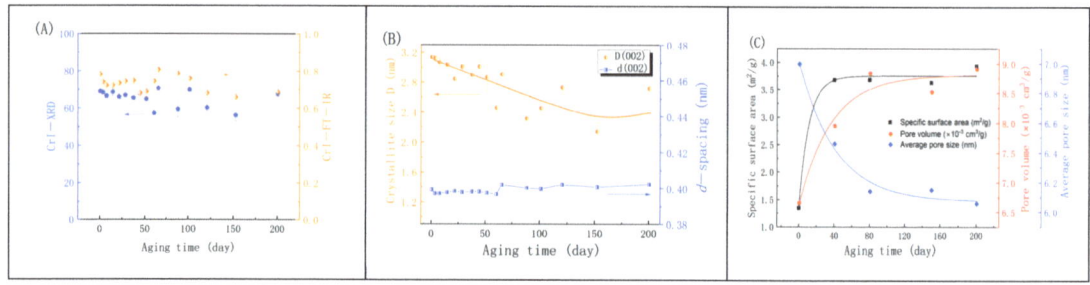

Figure 5. Variations in crystallinity index of cellulose (CrI) derived from XRD and FTIR (**A**), crystallite size D and d−spacing (**B**), and pore parameters (**C**) during thermal drying for bamboo paper.

Figure 5C illustrates the variation in pore parameters in paper fibers during the thermal drying process. In the early stages of aging, the specific surface area rapidly increases and stabilizes after 40 days, indicating that the fiber surface becomes rougher and more porous. The pore volume also shows a similar trend but with a slower growth rate, possibly due to the combined effect of pore size and quantity. In contrast, the average pore diameter gradually decreases as aging progresses, implying a contraction of the internal pore structure of the fibers. As the average pore size decreases while the pore volume increases, this indicates an increase in the number of pores. This may be attributed to the rapid breakage and degradation of cellulose molecular chains in the early stages of aging, leading to the formation of porous fibers. However, after 80 days of aging, the pore volume and average pore diameter of the fibers show a tendency to stabilize, indicating that the pore structure has reached a stable state. The average pore size does not further decrease, suggesting that the fibers maintain a relatively stable structure in the later stages of aging. Combining the previous discussion, this may be due to hemicellulose and lignin hindering cellulose aggregation and co-crystallization, maintaining the stability of the fiber skeleton structure. This phenomenon has also been reported in thermal treatment processes [43].

2.4. Aging Mechanisms

The aging behavior of bamboo paper is highly complex due to the presence of a certain proportion of hemicellulose and lignin in addition to cellulose in its chemical composition. As mentioned earlier, the mechanical properties exhibit three regions with respect to DP: the first plateau region, the rapid decline region, and the second plateau region (Figure 6). The DP of cellulose significantly impacts the mechanical properties of paper, and there is a similar trend between the cellulose content and the DP. It is evident that oxidative degradation dominates due to the significant decline in DP [23], leading to the speculation that fiber embrittlement may play a more significant role in the first plateau region than the deterioration of fiber–fiber bonds. It is commonly believed that the degradation of cellulose leads to chain breakage, reducing the inter-fiber bonding force and, thereby, decreasing the mechanical properties of paper. However, this study demonstrates that despite a significant decrease in cellulose DP, which would normally lead to brittleness in the initial plateau stage of fiber degradation, this brittleness does not significantly impair the mechanical performance of the paper. This may be attributed to the role of hydrogen bonds, as the strength of paper in the first plateau region remains unchanged. Additionally, the decrease in the WRV indicates the appearance of irreversible bonding in the paper, with some hornification occurring. The rearrangement of hydrogen bonds and hornification may compensate for the impact of cellulose chain breakage on mechanical strength. These hydrogen bonds, which are formed via free hydroxyl groups, create strong bonds within the fibers that are, to some extent, irreversible. The resulting multiple hydrogen bond structures are highly stable and not easily disrupted [48]. Additionally, the formation of ester bridges between cellulose molecular chains, which are covalent bonds and irreversible in water, contributes to hornification [49]. These various mechanisms may all contribute to the hornification process and potentially interact with each other. For instance, functional groups within the cellulose chains may engage in hydrogen bonds, ester bonds, ether bonds, and other types of bonding interactions. It is important to note that hornification is not entirely detrimental. This study suggests that moderate hornification can be strengthened by the hydrogen bonds between cellulose molecules, thereby enhancing the binding between fibers, helping to preserve the mechanical properties of paper, and preventing a sharp decline in the initial stage. It is important to note that the decrease in cellulose DP does not occur simultaneously with the decrease in cellulose content. During the aging process, cellulose chains may break, forming shorter chains, but this does not immediately reduce the total cellulose content. There may also be recombination of chains, keeping the total cellulose content relatively stable for a certain period of time.

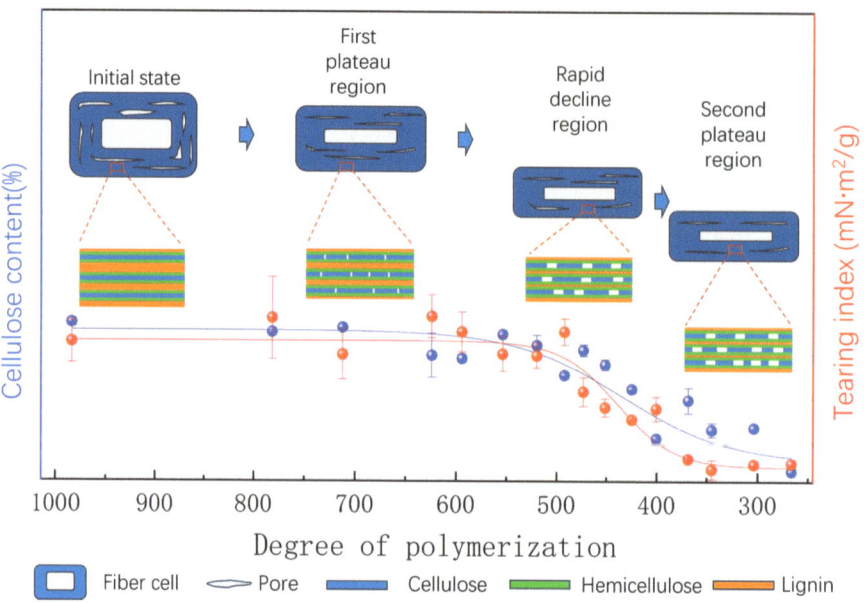

Figure 6. Schematic illustration of the structure and properties of traditional handmade bamboo paper during the aging process.

Once the cellulose DP falls below a critical value, a region of rapid decline appears. In this region, the most noticeable manifestation is further breakage of cellulose chains and the release of a large number of small molecules, leading to a rapid decrease in cellulose content and causing a series of adverse reactions, such as a decrease in paper acidity and a drastic decline in its mechanical properties. The degradation of cellulose or hemicellulose in paper can also generate volatile components that promote paper acidification. Cellulose in paper contains 1,4-β-glycosidic bonds, which are sensitive to acids. Its breakage leads to a decrease in the mechanical strength of the paper, as the reduction in DP affects the physical structure of the paper and the bonding between fibers. The structure of hemicellulose is more complex than that of cellulose, as it is composed of various sugars, including glucose, mannose, xylose, arabinose, and galactose. Under acidic conditions, hemicellulose also undergoes hydrolysis reactions, causing the glycosidic bonds to break and producing various monosaccharides [10].

The impact of hemicellulose and lignin on the aging behavior of bamboo paper is more positive than expected. Although the decrease in hemicellulose and lignin content is limited throughout the aging process, it is possible that their presence contributes to the appearance of the second plateau region.

Some studies suggest that lignin negatively impacts paper durability [14]. It plays a dual role in the degradation of cellulose—as an oxidation catalyst (source of radicals) and as an antioxidant [14,50]. In its role as an oxidative catalyst in an acidic paper, lignin generates free radicals that initiate and accelerate the oxidation process of cellulose. This process involves the hydroxyl (–OH) groups on the cellulose molecular chains, which can be oxidized into carbonyl (>C=O) and carboxyl (–COOH) groups, among other oxidation byproducts. This not only reduces the mechanical strength of the paper but also leads to changes in its chemical structure [14], further contributing to the yellowing of papers with high lignin content [51]. Recent studies have indicated that the lignin content does not affect the aging rate of paper produced at a neutral pH [52]. This is likely due to the cross-linking between lignin and carbohydrates effectively preventing the formation of hydrogen bonding in the fibers, both internally and externally, and reducing irreversible hornification [53].

The unchanged WRV indicates that fiber hornification does not deepen further, and the crystallinity does not continue to increase in this stage. Additionally, observations of the morphology of paper fibers show that after an initial significant contraction, there is no further contraction. Thus, the presence of hemicellulose and lignin limits the intermolecular interactions between cellulose molecules, slowing down the aggregation of the fiber filaments and maintaining the integrity of the fiber structure in the second plateau region.

This study only discusses the dry-heat aging behavior of handmade bamboo paper containing a certain amount of hemicellulose and lignin. However, different components of paper have their own characteristics in terms of chemical degradation processes under different aging conditions. Only through a comprehensive analysis of the structure and properties of paper under different aging conditions can a comprehensive conclusion of chemical degradation leading to changes in the mechanical properties of paper be provided. The chemical degradation of paper mainly occurs through the hydrolysis and oxidation of cellulose, which leads to embrittlement and failure of the paper [10]. A decrease in the DP reduces the effective mechanical properties of the fibers, ultimately leading to fiber embrittlement and loss of material integrity. In addition, the acidity, moisture content, and microstructural characteristics of the paper also affect its chemical and mechanical degradation behavior. Under different aging conditions, the mechanical behavior of paper varies. For example, high-humidity environments accelerate the hydrolysis of cellulose, causing the paper to embrittle more quickly. Conversely, in dry conditions, although the hydrolysis rate slows down, oxidation may become the primary degradation mechanism. Furthermore, the acidity of the paper affects its degradation rate, as an acidic environment accelerates the hydrolysis of cellulose, thus affecting the mechanical properties of the paper [7]. However, other protective materials may exhibit different aging behaviors from those of paper. For example, some polymer or synthetic materials may have different sensitivities to environmental conditions such as humidity and temperature or exhibit different mechanical behaviors during the aging process. These materials may have better aging resistance or exhibit more stable mechanical properties under specific conditions. However, their limitations lie in the fact that the glass transition temperature or that the melting point of polymers is higher than room temperature but lower than the aging temperature, making it difficult to evaluate their durability in relation to paper and further development of a comprehensive evaluation system is needed.

3. Materials and Methods

3.1. Materials

The bamboo paper used in this study was made from bitter bamboo and was crafted by hand using traditional craftsmanship. According to the "*Fenghua* City Chronicles", *Tang'ao* bamboo paper was first recorded in historical books in the ninth year of the *Zhengde* period of the Ming Dynasty (1514 AD), and it has a history of nearly 500 years. It can be used for the restoration of ancient books, and it has a thickness of 0.08 mm. The traditional bamboo papermaking process is roughly as follows: (1) harvesting bamboo for raw material; (2) air-drying the bamboo; (3) splitting the bamboo into strips; (4) soaking the strips in slaked lime; (5) cleaning the bamboo strips; (6) steaming for seven days; (7) pounding the bamboo into a pulp; (8) washing the paper pulp; (9) mixing with kiwi vine juice; (10) using a bamboo mat to form the paper; (11) pressing and drying; (12) removing the paper and trimming the edges. Copper ethylenediamine (CED) solution (Bis(ethylenediamine)copper(II) hydroxide solution (1 M in H_2O)) was purchased from Sigma-Aldrich (Shanghai) Trading Co., Ltd., Shanghai, China.

3.2. Accelerated Aging

The artificial aging treatment of handmade papers followed the Chinese standard GB/T464-2008 [54] (equivalent to ISO 5630-4:1986.MOD [55]) for dry-heat aging with the temperature set to 105 ± 2 °C and circa 0% RH. Sample collection was carried out according to a predetermined number of aging days. Sampling intervals were shorter in the early stages of aging and longer in the later stages. This condition was set to simulate the temperature effect during the natural aging process as closely as possible under controlled laboratory conditions in order to reflect the oxidation, fracture, and degradation of the paper itself.

3.3. Analysis of Chemical Components

The standard method of the United States Department of Energy was applied to quantitatively analyze the cellulose, hemicellulose, and lignin in the materials. The types and content of sugars in the hydrolyzed products were determined using high-performance liquid chromatography (HPLC). A 0.3 g sample was placed in a 10 mL crucible and mixed with 3 mL of 72% H_2SO_4, allowing it to swell for 60 min. Subsequently, 84 mL of distilled water was added to dilute the H_2SO_4 concentration to 4%. The mixture was reacted at 121 °C for 1 h and then filtered to obtain a solid residue (Klason lignin) and a filtrate. The acid-soluble lignin content in the filtrate was measured using Agilent 8453 ultraviolet–visible spectrophotometry (Agilent Technologies, Inc., Santa Clara, CA, USA) at 205 nm. The sugar content in the filtrate was analyzed via HPLC (LC-20AT) with an Aminex HPX-87H column (Bio-Rad Laboratories, Inc., Hercules, CA, USA) under the following conditions: a column temperature of 55 °C, a mobile phase of 5 mmol/L dilute H_2SO_4, and a flow rate of 0.6 mL/min.

3.4. Viscosity Determination

The DP value of the paper cellulose was measured using the viscosity method according to a report [1]. Paper samples were weighed and added to a plastic bottle with 10 mL of deionized water. After shaking for 30 min, 10 mL of CED (1 M in H_2O) solution was added. The plastic bottle was shaken evenly for 1 h at 25 °C until the paper specimen was dissolved completely. Then, the obtained solutions were transferred into a capillary viscometer, and the time of solution declining from the top to the bottom was recorded. Due to the high lignin content in the bamboo paper studied here, which was higher than 5%, the standard method for determining the intrinsic viscosity-related molecular weight of cellulose was no longer applicable. In this study, the intrinsic-viscosity-related molecular weight of cellulose was determined using a modified copper ethylenediamine method to determine the intrinsic viscosity and DP of handmade paper [56]. For handmade paper with a lignin content below 10%, the cellulose viscosity and DP can be obtained using the CED method based on Equation (1):

$$[\eta]_c = \frac{[\eta]_l}{1 - \chi\%}, \qquad (1)$$

where $[\eta]_c$ represents the intrinsic viscosity of cellulose without accounting for the lignin content (in mL/g); $[\eta]_l$ is the intrinsic viscosity of the model substance while accounting for lignin content (in mL/g); $\chi\%$ is the mass fraction of lignin.

The Martin empirical equation was used to calculate the DP of the paper samples (Equations (2)–(4)).

$$\eta_r = h_n * t_n, \qquad (2)$$

$$[\eta]_c = \eta_r / \rho, \qquad (3)$$

and

$$DP^{0.905} = 0.75[\eta]_c, \qquad (4)$$

where η_r is the relative viscosity of paper cellulose; h_n is the constant of the viscosimeter (0.0703 s^{-1}); t_n is the recorded time (s); $[\eta]_c$ is the intrinsic viscosity of paper cellulose; ρ is the concentration of the paper solution (g/mL).

3.5. Tests of Mechanical Properties

The tensile index of paper was measured according to GB/T 12914-2008 [57] at a constant elongation rate of 20 mm/min using a tensile strength tester (ZB-WLQ, Hangzhou Zhibang Automation Technology Co., Ltd., Hangzhou, China). The folding endurance was measured according to GB/T 475-2008 [58] with an MIT folding endurance tester (ZB-NZ135A, Hangzhou Zhibang Automation Technology Co., Ltd., Hangzhou, China). The tearing index was assessed according to GB/T 455-2002 [59] with a tearing strength tester (ZB-SL, Hangzhou Zhibang Automation Technology Co., Ltd., Hangzhou, China). The properties of the laboratory paper sheets were determined at a temperature of 23 °C and 50% relative humidity. The tensile index of paper is equal to the tensile strength divided by the base weight, and the tearing index is equal to the tear strength divided by the base weight.

3.6. Infrared Analysis

The attenuated total reflection Fourier transform infrared (ATR-FTIR) spectrum of the paper sample was determined on a Spectrum Two Spectrometer (PerkinElmer, Inc., Waltham, MA, USA) equipped with a diamond ATR detector. The scan scope was 4000–400 cm^{-1}. The original spectra were calibrated to eliminate the effects of radiation wavelength on the intensities of the absorption bands.

The hydrogen bond energy (E_H) was calculated using Equation (5) [60]:

$$E_H = \frac{1}{k}\left[\frac{v_0 - v}{v_0}\right], \qquad (5)$$

where $1/k = 2.625 \times 10^2$ kJ, v_0 is the frequency of the standard free hydroxyl group (3650 cm^{-1}), and v is the calculated frequency of the hydroxyl group.

The hydrogen bond length (R) was calculated using the Sederholm equation (Equation (6)):

$$v_0 - v = 4.43 \times 10^3 (2.84 - R), \qquad (6)$$

where v_0 is the stretching vibration frequency of a single hydroxyl group (3600 cm^{-1}), and v is the calculated frequency of the hydroxyl group.

3.7. Water Retention Value Measurement

The water retention value of the sample was tested as follows: 2 g of the sample was immersed in distilled water for 3 h, removed, and placed within a WRV tester to centrifuge for 30 min. The weight of the sample post-centrifugation was denoted as M_1. After that, the sample was transferred into an oven and dried at 105 °C for 4 h. After cooling, the weight of the sample, M_2, was measured, and the formula for calculating the water retention value was as follows:

$$WRV = \frac{M_1 - M_2}{M_2}. \qquad (7)$$

3.8. Chromaticity Test

The chromaticity of the paper was determined with an automatic colorimeter. To ensure accuracy and minimize potential errors, each sample was measured six times at different locations. The change in chromaticity ΔE was calculated via the following equation:

$$\Delta E = \sqrt{\Delta L^2 + \Delta a^2 + \Delta b^2}, \qquad (8)$$

where L, a, and b represent three different colorimetric coordinate values, respectively. L indicates the lightness, a indicates red–green, and b indicates yellow–blue. ΔL, Δa, and Δb are the differences between the corresponding values of different samples.

3.9. X-Ray Diffraction Measurement

The X-rays from a Cu tube operating at 30 kV and 10 mA were collected with an energy-dispersive detector that was able to resolve the Cu-Kα line (λ = 0.154184 nm). The X-ray source was a copper target bombarded with electrons. Scans were obtained from 5° to 40° 2θ using a step size of 0.05°.

The crystallinity index of cellulose was calculated from the XRD spectra using the following equation:

$$CrI = \frac{I_{200} - I_{AM}}{I_{200}} \times 100\%, \quad (9)$$

where I_{200} and I_{AM} are the scattering intensities from the diffraction intensity of the (200) lattice plane and the height of the minimum value between the (200) and the (110) peaks, respectively.

The *d*-spacing was calculated using Bragg's equation [10], and the crystallite sizes were calculated using the Scherrer equation [11]:

$$n\lambda = 2d\sin\theta \quad (10)$$

and

$$D = 0.9\lambda/(\beta\sin\theta) \quad (11)$$

where *n* is an integer; λ is the incident wavelength; *d* is the spacing between the planes in the atomic lattice; θ is the angle between the incident ray and the scattering planes; *D* is the crystallite size perpendicular to the plane; β is the full width at half-maximum in radians.

3.10. Low-Temperature Nitrogen Absorption

The low-temperature nitrogen adsorption measurement was conducted on the surface area and porosity analyzer (ASAP-2420, Micromeritics, Norcross, GA, USA). The sample (0.3–0.5 g, oven-dried weight) was then degassed at 120 °C for at least 8 h. The N_2 adsorption/desorption isotherms were analyzed at 77 K, and the specific surface area was calculated using the BET method. The average pore size and the pore volume were calculated using the BJH method.

4. Conclusions

This study explores the structural and performance changes in traditional Chinese bamboo paper during the process of degradation through dry-heat aging. The research reveals that the DP of cellulose undergoes three phases: an initial plateau phase, a rapid decline phase, and a second plateau phase. A critical performance threshold is observed when the DP ranges from 600 to 400, marking a shift from a balanced or slightly decreasing trend in the initial plateau phase to a sharp decline. There is a distinct correlation between the decrease in cellulose content in paper and the deterioration of certain paper properties. This study also discusses for the first time that the formation of the second plateau phase may be due to the presence of hemicellulose and lignin, which hinder further aggregation of cellulose and maintain structural stability, providing the paper with some strength even after 200 days of dry-heat aging. While there are limitations in this study, obtaining detailed insights into hydrogen bond rearrangements and chemical group bonding would enhance explanations from a chemical and structural perspective, providing a deeper understanding of the structural and performance changes in the bamboo paper during degradation through dry-heat aging. This research can provide valuable guidance for traditional papermaking practices and the preservation of ancient books. For papers that contain a certain amount of hemicellulose and lignin, these components may play a role in maintaining structural stability during longer-term preservation. Meanwhile, selecting papers with a higher cellulose DP can prolong the duration of the initial plateau phase, thereby extending the preservation time of ancient books. On the other hand, this study provides multidimensional considerations for the development of materials for the preservation of ancient books. After implementation, it is important to examine whether the degradation and hornification of paper are inhibited and whether the maintenance of apparent performance affects the extent of internal degradation. The limitations of this study include the need to systematically compare the degradation behavior of bamboo paper under different humidity conditions, UV exposure, or exposure to different air pollutants and to establish dynamic or thermodynamic models that describe the relationship between cellulose degradation and accelerated aging parameters under the influence of hemicellulose and lignin. This would provide a scientific basis for a deeper understanding of aging mechanisms and the restoration and preservation of traditional ancient books.

Author Contributions: Conceptualization, X.J. and P.L.; methodology, Z.Z., K.Z., Y.W., Y.Z. and P.L.; validation, Z.Z., K.Z., Y.X. and Z.L.; formal analysis, X.J.; investigation, Z.Z., K.Z., Y.X., Z.L., Y.W., Y.Z., X.J. and P.L.; resources, X.J. and P.L.; data curation, Z.Z., K.Z. and Y.W.; writing—original draft preparation, Z.Z. and K.Z.; writing—review and editing, Z.Z., K.Z., X.J. and P.L. All authors have read and agreed to the published version of the manuscript.

Funding: This research was funded by the Natural Science Foundation of Shanghai, China (21ZR1405100); the Foundation of the State Key Laboratory of Biobased Material and Green Papermaking, Qilu University of Technology, Shandong Academy of Sciences, China (GZKF202210); and the Humanities and Social Sciences Foundation of the Ministry of Education of China (21YJCZH089).

Institutional Review Board Statement: Not applicable.

Informed Consent Statement: Not applicable.

Data Availability Statement: Data are contained within the article.

Conflicts of Interest: The authors declare no conflicts of interest.

References

1. Zhang, X.; Liu, P.; Yan, Y.; Yao, J.; Tang, Y.; Yang, Y. Degradation of Chinese handmade papers with different fiber raw materials on molecular and supramolecular structures. *Polym. Degrad. Stab.* **2023**, *211*, 110330. [CrossRef]
2. Hubbe, M.A.; Bowden, C. Handmade paper: A review of its history, craft, and science. *BioResources* **2009**, *4*, 1736–1792. [CrossRef]
3. Jain, P.; Gupta, C. A sustainable journey of handmade paper from past to present: A review. *Probl. Ekorozwoju* **2021**, *16*, 234–244. [CrossRef]
4. Chen, G.; Katsumata, K.S.; Inaba, M. Traditional Chinese Papers, their Properties and Permanence. *Restaurator. Int. J. Preserv. Libr. Arch. Mater.* **2003**, *24*, 135–144. [CrossRef]
5. Fan, Y.; Han, B.; Ge, M.; Jiang, R.; Shi, J. Development process and reserch status of handmade bamboo paper. *World Bamboo Ratt.* **2024**, *22*, 100–106.
6. Gang, C. Current Situation and Questions of Archives and Ancient Books Restoration in Bamboo Paper. *Arch. Sci. Study* **2012**, 80–84. Available online: https://link.cnki.net/doi/10.16065/j.cnki.issn1002-1620.2012.01.019 (accessed on 25 May 2012).
7. Baty, J.W.; Maitland, C.L.; Minter, W.; Hubbe, M.A.; Jordan-Mowery, S.K. Deacidification for the conservation and preservation of paper-based works: A review. *BioResources* **2010**, *5*, 1955–2023. [CrossRef]
8. Ahn, K.; Hennniges, U.; Banik, G.; Potthast, A. Is cellulose degradation due to β-elimination processes a threat in mass deacidification of library books? *Cellulose* **2012**, *19*, 1149–1159. [CrossRef]
9. Ahn, K.; Rosenau, T.; Potthast, A. The influence of alkaline reserve on the aging behavior of book papers. *Cellulose* **2013**, *20*, 1989–2001. [CrossRef]
10. Xu, Z.; Yueer, Y.; Jingjing, Y.; Shutong, J.; Yi, T. Chemistry directs the conservation of paper cultural relics. *Polym. Degrad. Stab.* **2023**, *207*, 110228. [CrossRef]
11. Zervos, S.; Moropoulou, A. Methodology and criteria for the evaluation of paper conservation interventions: A literature review. *Restaurator* **2006**, *27*, 219–274. [CrossRef]
12. Liu, Y.; Zhou, Y.; Lu, J. Exploring the relationship between air pollution and meteorological conditions in China under environmental governance. *Sci. Rep.* **2020**, *10*, 14518. [CrossRef] [PubMed]
13. Łojewska, J.; Miśkowiec, P.; Łojewski, T.; Proniewicz, L.M. Cellulose oxidative and hydrolytic degradation: In situ FTIR approach. *Polym. Degrad. Stab.* **2005**, *88*, 512–520. [CrossRef]
14. Łojewski, T.; Zięba, K.; Knapik, A.; Bagniuk, J.; Lubańska, A.; Łojewska, J. Evaluating paper degradation progress. Cross-linking between chromatographic, spectroscopic and chemical results. *Appl. Phys. A* **2010**, *100*, 809–821. [CrossRef]
15. Piantanida, G.; Bicchieri, M.; Coluzza, C. Atomic force microscopy characterization of the ageing of pure cellulose paper. *Polymer* **2005**, *46*, 12313–12321. [CrossRef]
16. Lv, S. A research on the characteristics of Chinese ancient paper aging. *Anc. Books Conserv. Study* **2020**, *2*, 73–88.
17. Xu, W. Research on the aging performance of paper used for the restoration of calligraphy and painting cultural relics. *China Cult. Herit. Sci. Res.* **2021**, *2*, 68–72.

18. Tian, Z.; Long, K.; Ren, S.; Zhang, M. A study on in study on influence of storage environment on paper properties. *China Pulp Pap. Ind.* **2016**, *37*, 31–33.
19. Chen, B.; Tan, J.; Huang, J.; Lu, Y.; Gu, P.; Han, Y.; Ding, Y. Research on the aging—Resistance properties of four kinds of Fuyang bamboo paper. *J. For. Eng.* **2021**, *6*, 121–126. [CrossRef]
20. Tan, J.; Lu, Y.; Fu, X.; Chen, B.; Ding, Y. Effects of micro-structural changes on properties of aged bamboo paper. *Chin. Sci. Bull.* **2022**, *67*, 4429–4438. [CrossRef]
21. Chen, B.; Tan, J.; Fu, X.; Lu, Y.; Zhu, Y.; Huang, J.; Di, Y.; Ding, Y. Study on Pyrolysis Characteristics of Bamboo Paper After Aging and Quantitative Evaluation of Its Aging Degree. *Mater. Rep.* **2022**, *36*, 213–217. [CrossRef]
22. Yi, X.; Li, Y.; Lei, X. Study on Performance Difference of Handmade Bamboo Paper Between Traditional Uncooked Process and Cooking Process. *Trans. China Pulp Pap.* **2022**, *37*, 78–85. [CrossRef]
23. Vibert, C.; Dupont, A.-L.; Dirrenberger, J.; Passas, R.; Ricard, D.; Fayolle, B. Relationship between chemical and mechanical degradation of aged paper: Fibre versus fibre–fibre bonds. *Cellulose* **2024**, *31*, 1855–1873. [CrossRef]
24. Stephens, C.H.; Whitmore, P.M. Comparison of the degradation behavior of cotton, linen, and kozo papers. *Cellulose* **2013**, *20*, 1099–1108. [CrossRef]
25. KATO, K.L.; CAMERON, R.E. Structure–Property Relationships in Thermally Aged Cellulose Fibers and Paper. *J. Appl. Polym. Sci.* **1999**, *74*, 1465–1477. [CrossRef]
26. Vizárová, K.; Kirschnerová, S.; Kačík, F.; Briškárová, A.; Šutý, Š.; Katuščák, S. Relationship between the decrease of degree of polymerization of cellulose and the loss of groundwood pulp paper mechanical properties during accelerated ageing. *Chem. Pap.* **2012**, *66*, 1124–1129. [CrossRef]
27. Ding, H.Z.; Wang, Z.D. On the degradation evolution equations of cellulose. *Cellulose* **2008**, *15*, 205–224. [CrossRef]
28. Zou, X.; Uesaka, T.; Gurnagul, N. Prediction of paper permanence by accelerated aging I. Kinetic analysis of the aging process. *Cellulose* **1996**, *3*, 243–267. [CrossRef]
29. Zou, X.; Gurnagul, N.; Uesaka, T.; Bouchard, J. Accelerated aging of papers of pure cellulose: Mechanism of cellulose degradation and paper embrittlement. *Polym. Degrad. Stab.* **1994**, *43*, 393–402. [CrossRef]
30. Gurnagul, N.; Howard, R.C.; Zou, X.; Uesaka, T.; Page, D.H. Mechanical permanence of paper: A literature review. *J. Pulp Pap. Sci.* **1993**, *19*, 160–166. [CrossRef]
31. Jin, C.; Wu, C.; Liu, P.; Yu, H.; Yang, Y.; Zhang, H. Kinetics of cellulose degradation in bamboo paper. *Nord. Pulp Pap. Res. J.* **2022**, *37*, 480–488. [CrossRef]
32. Jeong, M.-J.; Kang, K.-Y.; Bacher, M.; Kim, H.-J.; Jo, B.-M.; Potthast, A. Deterioration of ancient cellulose paper, Hanji: Evaluation of paper permanence. *Cellulose* **2014**, *21*, 4621–4632. [CrossRef]
33. Fayolle, B.; Richaud, E.; Colin, X.; Verdu, J. Review: Degradation-induced embrittlement in semi-crystalline polymers having their amorphous phase in rubbery state. *J. Mater. Sci.* **2008**, *43*, 6999–7012. [CrossRef]
34. Klemm, D.; Heublein, B.; Fink, H.-P.; Bohn, A. Cellulose: Fascinating Biopolymer and Sustainable Raw Material. *Angew. Chem. Int. Ed.* **2005**, *44*, 3358–3393. [CrossRef]
35. Yi, T.; Zhao, H.; Mo, Q.; Pan, D.; Liu, Y.; Huang, L.; Xu, H.; Hu, B.; Song, H. From Cellulose to Cellulose Nanofibrils—A Comprehensive Review of the Preparation and Modification of Cellulose Nanofibrils. *Materials* **2020**, *13*, 5062. [CrossRef] [PubMed]
36. Park, S.; Baker, J.O.; Himmel, M.E.; Parilla, P.A.; Johnson, D.K. Cellulose crystallinity index: Measurement techniques and their impact on interpreting cellulase performance. *Biotechnol. Biofuels* **2010**, *3*, 10. [CrossRef]
37. Medhekar, N.V.; Ramasubramaniam, A.; Ruoff, R.S.; Shenoy, V.B. Hydrogen Bond Networks in Graphene Oxide Composite Paper: Structure and Mechanical Properties. *ACS Nano* **2010**, *4*, 2300–2306. [CrossRef]
38. Toba, K.; Yamamoto, H.; Yoshida, M. Crystallization of cellulose microfibrils in wood cell wall by repeated dry-and-wet treatment, using X-ray diffraction technique. *Cellulose* **2013**, *20*, 633–643. [CrossRef]
39. Lin, Q.; Huang, Y.; Yu, W. An in-depth study of molecular and supramolecular structures of bamboo cellulose upon heat treatment. *Carbohydr. Polym.* **2020**, *241*, 116412. [CrossRef]
40. Leng, E.; Gong, X.; Zhang, Y.; Xu, M. Progress of cellulose pyrolysis mechanism: Cellulose evolution based on intermediate cellulose. *CIESC J.* **2018**, *69*, 239–248.
41. Zervos, S. Characterization of changes induced by ageing to the microstructure of pure cellulose paper by use of a gravimetric water vapour adsorption technique. *Cellulose* **2007**, *14*, 375–384. [CrossRef]
42. Kato, K.L.; Cameron, R.E. A Review of the Relationship Between Thermally-Accelerated Ageing of Paper and Hornification. *Cellulose* **1999**, *6*, 23–40. [CrossRef]
43. Mo, W.; Chen, K.; Yang, X.; Kong, F.; Liu, J.; Li, B. Elucidating the hornification mechanism of cellulosic fibers during the process of thermal drying. *Carbohydr. Polym.* **2022**, *289*, 119434. [CrossRef]
44. Heinze, T. Cellulose: Structure and Properties. In *Cellulose Chemistry and Properties: Fibers, Nanocelluloses and Advanced Materials*; Rojas, O.J., Ed.; Springer International Publishing: Cham, Switzerland, 2016; pp. 1–52. [CrossRef]
45. Yao, J.; Zhang, R.; Luo, C.; Yan, Y.; Bi, N.; Tang, Y. Deterioration of Kaihua handmade paper: Evolution of molecular, supramolecular and macroscopic structures. *Polym. Degrad. Stab.* **2022**, *195*, 109773. [CrossRef]
46. Wan, J.; Wang, Y.; Xiao, Q. Effects of hemicellulose removal on cellulose fiber structure and recycling characteristics of eucalyptus pulp. *Bioresour. Technol.* **2010**, *101*, 4577–4583. [CrossRef]

47. Sjöstrand, B.; Karlsson, C.-A.; Barbier, C.; Henriksson, G. Hornification in Commercial Chemical Pulps: Dependence on Water Removal and Hornification Mechanisms. *BioResources* **2023**, *18*, 3856–3869. [CrossRef]
48. Zhou, M.; Chen, D.; Chen, Q.; Chen, P.; Song, G.; Chang, C. Reversible Surface Engineering of Cellulose Elementary Fibrils: From Ultralong Nanocelluloses to Advanced Cellulosic Materials. *Adv. Mater.* **2024**, *36*, 2312220. [CrossRef] [PubMed]
49. Wohlert, M.; Benselfelt, T.; Wågberg, L.; Furó, I.; Berglund, L.A.; Wohlert, J. Cellulose and the role of hydrogen bonds: Not in charge of everything. *Cellulose* **2022**, *29*, 1–23. [CrossRef]
50. Barclay, L.R.C.; Xi, F.; Norris, J.Q. Antioxidant Properties of Phenolic Lignin Model Compounds. *J. Wood Chem. Technol.* **1997**, *17*, 73–90. [CrossRef]
51. Rychlý, J.; Matisová-Rychlá, L.; Bukovský, V.; Pleteníková, M.; Vrška, M. The Progress of Ageing of Lignin-containing Paper Induced by Light and its Relation to Chemiluminescence—Temperature Runs. *Macromol. Symp.* **2005**, *231*, 178–192. [CrossRef]
52. Małachowska, E.; Dubowik, M.; Boruszewski, P.; Łojewska, J.; Przybysz, P. Influence of lignin content in cellulose pulp on paper durability. *Sci. Rep.* **2020**, *10*, 19998. [CrossRef] [PubMed]
53. Fu, H.; Gao, W.; Wang, B.; Zeng, J.; Cheng, Z.; Xu, J.; Chen, K. Effect of lignin content on the microstructural characteristics of lignocellulose nanofibrils. *Cellulose* **2020**, *27*, 1327–1340. [CrossRef]
54. *GB/T 464-2008*; Paper and Board—Accelerated Aging—Dry Heat Treatment. General Administration of Quality Supervision, Inspection and Quarantine of the People's Republic of China, China National Standardization Management Committee: Beijing, China, 2008.
55. *ISO 5630-4:1986.MOD*; Paper and Board; Accelerated Ageing; Part 4: Dry Heat Treatment at 120 or 150 Degrees. International Organization for Standardization (ISO): Geneva, Switzerland, 1986.
56. Liu, P.; Zhang, R.; Jiang, Y.; Liu, J.; Sun, M.; Yan, Y.; Tang, Y. Studies on the applicability of a modified copper ethylenediamine method for the determination of intrinsic viscosity and degree of polymerization of handmade paper. *Sci. Conserv. Archaeol.* **2019**, *31*, 1–5. [CrossRef]
57. *GB/T 12914-2008*; Paper and Board—Determination of Tensile Properties. General Administration of Quality Supervision, Inspection and Quarantine of the People's Republic of China, China National Standardization Management Committee: Beijing, China, 2008.
58. *GB/T 457-2008*; Paper and Board—Determination of Folding Endurance. General Administration of Quality Supervision, Inspection and Quarantine of the People's Republic of China, China National Standardization Management Committee: Beijing, China, 2008.
59. *GB/T 455-2002*; Pulp and Board Determination of Tearing Resistance. General Administration of Quality Supervision, Inspection and Quarantine of the People's Republic of China: Beijing, China, 2002.
60. Turki, A.; El Oudiani, A.; Msahli, S.; Sakli, F. Investigation of OH bond energy for chemically treated alfa fibers. *Carbohydr. Polym.* **2018**, *186*, 226–235. [CrossRef] [PubMed]

Disclaimer/Publisher's Note: The statements, opinions and data contained in all publications are solely those of the individual author(s) and contributor(s) and not of MDPI and/or the editor(s). MDPI and/or the editor(s) disclaim responsibility for any injury to people or property resulting from any ideas, methods, instructions or products referred to in the content.

Article

Influence of Relative Humidity on the Mechanical Properties of Palm Leaf Manuscripts: Short-Term Effects and Long-Term Aging

Wenjie Zhang [1], Shan Wang [2] and Hong Guo [1,*]

[1] Key Laboratory of Archaeomaterials and Conservation, Institute of Cultural Heritage and History of Science & Technology, University of Science and Technology Beijing, Ministry of Education, Beijing 100083, China; d202310769@xs.ustb.edu.cn
[2] Chinese Academy of Cultural Heritage, Beijing 100029, China; cnicpbj@gmail.com
* Correspondence: guohong@ustb.edu.cn

Abstract: Palm leaf manuscripts are a valuable part of world cultural heritage. Studying the mechanical properties of palm leaf manuscripts and their changes due to environmental influences is of great significance for understanding the material characteristics, aging mechanisms, and preventive conservation of these manuscripts. This study used dynamic vapor sorption (DVS) and a thermomechanical analyzer (TMA) to investigate the changes to the mechanical properties of palm leaf manuscripts in response to different relative humidity conditions and different time periods. The short-term study results show that exposure to varying relative humidities leads to changes in the equilibrium moisture content (EMC) of palm leaf manuscripts, causing the bending strength of the samples to decrease significantly with increasing humidity. The bending modulus initially increases and then decreases as the humidity increases. Moreover, the greater the desorption hysteresis of the samples, the more pronounced the changes to the mechanical properties. Therefore, a stable environment in terms of humidity can prevent changes in the mechanical properties of palm leaf manuscripts, thereby preventing the onset of degradation. The results of the long-term aging studies indicate that prolonged exposure to either very dry or very humid conditions greatly affects the mechanical properties of palm leaf manuscripts, which is detrimental to their preservation. The samples kept at 50% RH did not exhibit significant signs of deterioration, with no notable changes in their mechanical properties or chemical structure. This suggests that 50% RH is a relatively optimal humidity condition for the preservation of palm leaf manuscripts.

Keywords: palm leaf manuscripts; mechanical properties; dynamic vapor sorption; thermomechanical analysis; simulated aging experiment; preventive conservation

Citation: Zhang, W.; Wang, S.; Guo, H. Influence of Relative Humidity on the Mechanical Properties of Palm Leaf Manuscripts: Short-Term Effects and Long-Term Aging. *Molecules* 2024, 29, 5644. https://doi.org/10.3390/molecules29235644

Academic Editors: Carmelo Corsaro, Yi Tang, Yueer Yan and Yuliang Yang

Received: 26 October 2024
Revised: 27 November 2024
Accepted: 27 November 2024
Published: 28 November 2024

Copyright: © 2024 by the authors. Licensee MDPI, Basel, Switzerland. This article is an open access article distributed under the terms and conditions of the Creative Commons Attribution (CC BY) license (https:// creativecommons.org/licenses/by/ 4.0/).

1. Introduction

Before the widespread use of paper, palm leaf manuscripts served as a popular literary medium in South Asia and Southeast Asia, carrying rich cultural significance. These manuscripts not only recorded knowledge across various fields, such as history, literature, philosophy, art, and science, but also held an important place in Buddhist culture and religious teachings [1–3]. Given their immense historical and cultural value, palm leaf manuscripts are recognized as a precious part of world cultural heritage, making their preservation and preventive conservation especially important.

At present, significant progress has been made in regard to the collection and preservation of palm leaf manuscripts, as well as in research related to cataloging and text interpretation [4–6]. Some scholars have conducted systematic analyses of the materials [7–10] and craftsmanship [11,12] used in the production of palm leaf manuscripts, completed preliminary assessments of degradation [13], and classified common types of damage [14,15]. Based on these studies, conservation professionals have implemented a series of measures [16,17], making important contributions to the preservation and restoration of palm leaf manuscripts. These studies provide foundational data for the scientific understanding and preventive conservation of palm leaf manuscripts and offer valuable references for selecting appropriate preservation environments.

Although there have been some advancements in the research on production techniques, material analysis, and preservation and restoration methods for palm leaf manuscripts, studies on the effects and mechanisms of environmental factors on their aging remain insufficient. Several studies have preliminarily explored the impact of environmental conditions on the physical properties of palm leaf manuscripts. For example, dry environments may reduce the mechanical strength of palm leaf manuscripts [12,18], while excessively high humidity significantly enhances their hygroscopicity [19]. Additionally, in conservation research on other cellulose-based materials like paper and wood, extensive studies have investigated the aging effects and mechanisms of various environmental factors. For instance, high temperatures accelerate the thermal oxidative degradation of cellulose and hemicellulose, resulting in a decline in their mechanical properties [20,21]. Similarly, exposure to high-energy ultraviolet radiation breaks the molecular chains of cellulose, leading to mechanical deterioration and material disintegration [22]. This radiation also induces photochemical reactions in components such as lignin, causing discoloration of the material [23]. Furthermore, acidic gases, like sulfur compounds and nitrogen oxides, react with moisture in the material, triggering acidic hydrolysis [24]. The accumulation of dust directly alters the color and surface properties of the material [25]. These findings offer valuable insights into the preservation of palm leaf manuscripts, particularly for preventive conservation. Notably, as both a literary medium and a cultural heritage artifact, palm leaf manuscripts are typically stored indoors, in libraries or study rooms, thereby largely shielding them from external temperature fluctuations, light exposure, and atmospheric pollutants. Consequently, humidity, which tends to fluctuate more readily, becomes the primary environmental factor to consider in the preservation of palm leaf manuscripts.

Humidity is closely linked to the mechanical properties of palm leaf manuscripts, which directly impact their preservation and restoration processes. Variations in humidity can cause physical deformation of the manuscripts, leading to permanent structural damage [26], while higher humidity levels promote the growth and proliferation of microorganisms [27]. These factors can affect the mechanical properties of palm leaf manuscripts [12] and accelerate material deterioration. Moreover, whether short-term and long-term exposure to the same relative humidity has different effects on the mechanical properties of the manuscripts remains a key question that could lead to the adoption of different approaches to preservation and restoration. Therefore, studying the mechanical properties of palm leaf manuscripts and how they change under the influence of relative humidity is essential for a deeper understanding of their intrinsic characteristics and aging mechanisms. This research is crucial for developing effective restoration techniques and preventive conservation measures.

Dynamic vapor sorption (DVS) can be used to study the hygroscopic behavior of materials in response to different relative humidity conditions. It has shown great potential in regard to evaluating the hygroscopicity of palm leaf manuscripts [28] and other cellulose-based artifacts, such as wood and paper [29–31]. Unlike traditional mechanical testing equipment, such as universal testing machines, which require larger sample sizes, a thermomechanical analyzer (TMA) offers advantages in regard to testing mechanical properties using small sample sizes, with high precision and good repeatability [32,33], making it particularly suitable for fragile and precious organic artifacts, like palm leaf manuscripts. However, a TMA lacks intrinsic humidity control capabilities, and traditional humidity controlled equipment, such as desiccators and constant temperature humidity chambers, cannot be directly integrated with a TMA. This setup creates a lag between humidity conditioning and mechanical testing, making it challenging to capture the immediate effects of humidity on the material's mechanical properties. Furthermore, fluctuations in humidity make it difficult to accurately measure the subtle short-term mechanical responses of the material. To overcome these limitations and enable synchronized humidity control and mechanical testing, ensuring real-time accuracy of the mechanical data in response to varying humidity conditions, this study innovatively employs a custom-made Modular Humidity Generator (MHG) connected to the TMA. This method allows precise humidity adjustment and captures the mechanical response of palm leaf manuscripts in diverse humidity environments, providing a scientific basis for studying humidity's impact on the material's mechanical properties.

Using raw palm leaf manuscripts as research samples, this study examined the short-term effects of varying humidity on their mechanical properties, as well as the long-term aging effects. This approach aims to deepen our understanding of how environmental factors influence the lifespan of palm leaf manuscripts and provide data that support research on their degradation and preventive conservation.

2. Results and Discussion

2.1. Hygroscopicity of the Samples

The relationship between the equilibrium moisture content (EMC) of the samples (PLSs) and the environmental relative humidity (RH) is described by an isothermal adsorption curve [34]. The isothermal adsorption curve of the samples is shown in Figure 1a. The curve follows the trend of an IUPAC Type II adsorption isotherm [35]. During the adsorption phase (ad), the EMC of the samples increases with the increasing relative humidity, showing a gradual rise followed by a sharp increase, which is characteristic of the hygroscopic behavior of cellulose-based materials. At 95% RH, the EMC of the samples reached 25.01%. During the desorption phase (de), the EMC decreases as the relative humidity drops, but remains higher than in the adsorption phase, indicating a certain level of desorption hysteresis.

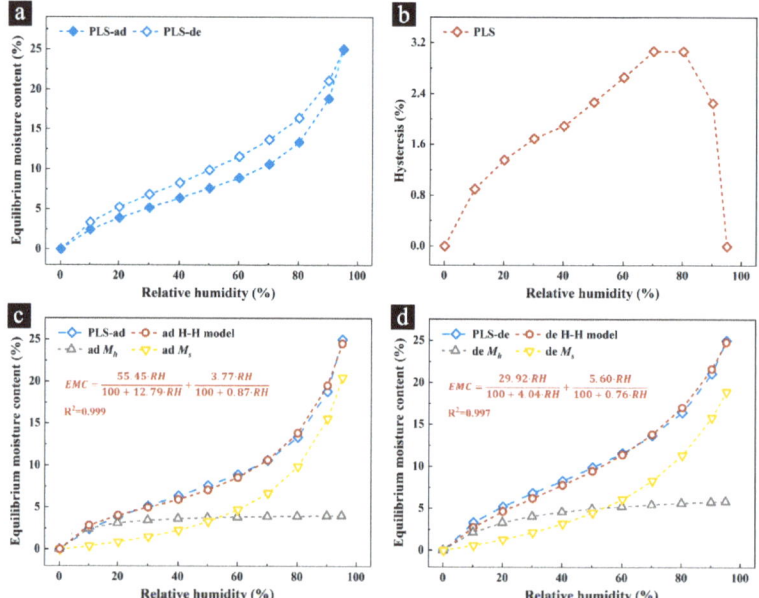

Figure 1. The isothermal adsorption curve of the samples (**a**); the hysteresis curve (**b**); the Hailwood–Horrobin (H–H) model fitting results for the adsorption curve (**c**); and the desorption curve (**d**) of the samples (at 25 °C, 0% RH to 95% RH).

Desorption hysteresis refers to the lag phenomenon that occurs when the behavior of a substance during adsorption and desorption is not identical. This hysteresis is expressed by the hysteresis value, which represents the difference in the equilibrium moisture content (EMC) between the desorption and adsorption phases. The hysteresis curve of the sample is shown in Figure 1b. The hysteresis value of the samples initially increases and then decreases with increasing relative humidity, reaching a maximum value of 3.08% at 70% RH.

To further study the adsorption behavior of the samples, the isothermal adsorption curves were fitted to the Hailwood–Horrobin (H–H) model. Table 1 lists the calculated parameters for each model. The high coefficient of determination ($R^2 > 0.99$) confirms the validity of the model.

Table 1. The H–H model parameters for the samples.

Process	w	k_1	k_2	R^2
ad	415.087	14.717	0.869	0.999
de	297.972	9.740	0.746	0.997

In the table, ad represents the adsorption process; de represents the desorption process; w is the molecular weight of the wood at each adsorption site; and k_1 and k_2 are the equilibrium constants related to monolayer and multilayer adsorption, respectively.

Using the calculated model parameters, the adsorption curve (Figure 1c) and desorption curve (Figure 1d) of the samples were plotted. The results indicate that the H–H model effectively explains the adsorption and desorption processes of the samples. According to the H–H model, the total adsorbed moisture content in a material can be divided into monolayer moisture content (M_h) and multilayer moisture content (M_s) [36]. Typically, in low humidity environments (0% to 40% RH), the adsorption of moisture by cellulose materials is primarily driven by monolayer adsorption. As the relative humidity increases and the monolayer adsorption sites are occupied, monolayer adsorption gradually stabilizes and multilayer adsorption plays an increasingly important role in the moisture adsorption process. This pattern can be observed in both the adsorption and desorption phases of the samples.

During desorption, the M_h value of the sample increases compared to the adsorption phase, while the M_s value remains unchanged. This suggests that the observed hysteresis may be due to changes in the level of monolayer adsorption between the two processes.

2.2. Short-Term Effects of Relative Humidity on the Mechanical Properties of the Samples

2.2.1. Adsorption Process

The TMA-MHG system was used to simulate and measure the changes in the mechanical properties of the samples as the humidity increased, represented by the flexural strength (Figure 2a) and flexural modulus (Figure 2b). The results indicate that the TMA is a viable method for measuring the mechanical properties of precious and relatively fragile organic materials, such as palm leaf manuscripts. The flexural strength of the samples decreased significantly with increasing humidity, with the highest flexural strength observed at 10% RH (44.86 MPa), while at 90% RH, the flexural strength dropped to only 22.03 MPa. The flexural modulus exhibited a different trend. The highest flexural modulus (657.26 MPa) was observed at 50% RH, while both low and high humidity conditions resulted in a reduced flexural modulus: 380.38 MPa at 10% RH and 208.06 MPa at 90% RH.

This variation aligns with the characteristics of cellulose-based materials. At a low moisture content, the hydrogen bonding between cellulose molecules is stronger, leading to tightly bound fiber bundles and higher overall mechanical strength. However, dry conditions can also increase the brittleness of the material, reducing its flexibility and, thus, its flexural strength or stiffness [37]. In humid environments, cellulose materials absorb large amounts of water and the absorbed water molecules infiltrate the cellulose structure, forming hydrogen bonds with some cellulose molecules. This weakens the intermolecular forces between cellulose chains, significantly reducing the material's overall mechanical strength. As the moisture content increases, the material becomes more flexible and stiffness decreases, as reflected by the simultaneous reduction in both the flexural strength and flexural modulus [38,39]. Similar phenomena have been observed in regard to other cellulose-based materials, such as wood and paper [37–39].

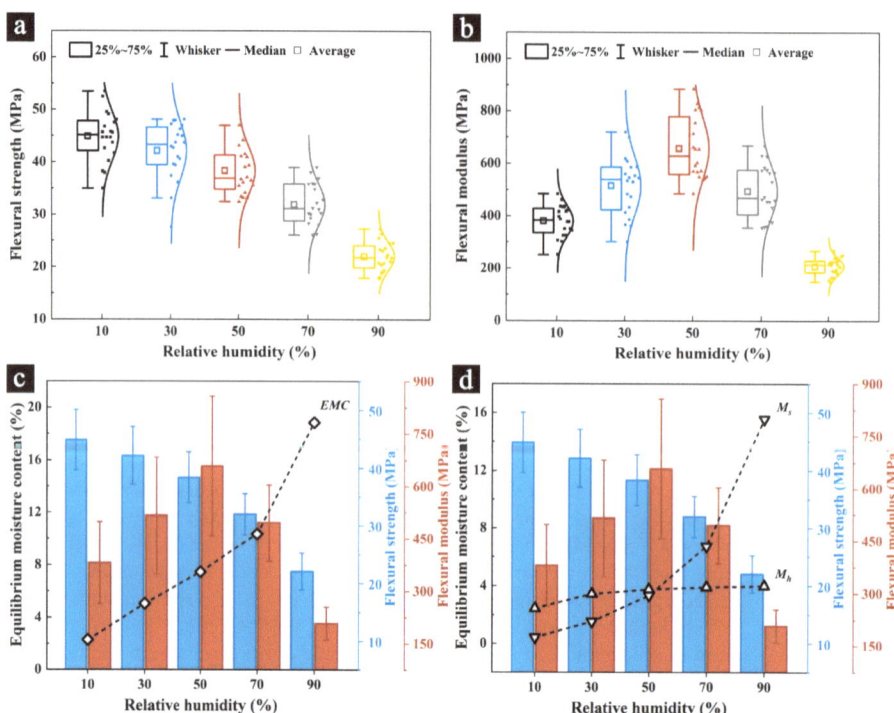

Figure 2. The flexural strength (**a**) and flexural modulus (**b**) of the samples at different humidity levels during the adsorption process, along with the correlations between the flexural strength and modulus with the EMC (**c**), and M_h and M_s (**d**).

The isothermal moisture adsorption curve of the samples (Figure 2c) and the variation of M_h and M_s with humidity (Figure 2d) further confirm the changes in the mechanical properties of the samples. At a relative humidity of 10% RH, the EMC of the sample was 2.43%, corresponding to the highest flexural strength and a lower flexural modulus. As the humidity increased, the EMC of the sample rose significantly, while the flexural strength gradually decreased and the flexural modulus first increased and then decreased. At 90% RH, the EMC reached 18.85%, at which point both the flexural strength and modulus were at their lowest values.

In low humidity environments (0–40% RH), the EMC of the sample was primarily composed of monolayer moisture content (M_h), resulting in less pronounced changes in the flexural strength. As the humidity increased, M_h stabilized, while the multilayer moisture content (M_s) increased significantly. The M_s value typically reflects the material's ability to adsorb moisture through physical adsorption and capillary action [40]. This indicates that the physical adsorption and capillary action of the sample increased markedly with increasing humidity, leading to an increase in the EMC. The water molecules formed hydrogen bonds with the cellulose molecules, weakening the intermolecular forces between the cellulose fibers. This phenomenon manifested macroscopically as a reduction in the mechanical strength and stiffness of the sample [38,39].

2.2.2. Desorption Process

The TMA-MHG system was used to simulate and measure the changes in the mechanical properties of the samples as the humidity decreased, represented by the flexural strength (Figure 3a) and flexural modulus (Figure 3b). The results indicate that the TMA is a viable method for measuring the mechanical properties of precious and relatively fragile organic materials, such as palm leaf manuscripts. The trend in the changes to both the flexural strength and flexural modulus during the desorption process was consistent with the adsorption process: the flexural strength significantly decreased as the humidity increased and the flexural modulus decreased in response to both low and high humidity conditions.

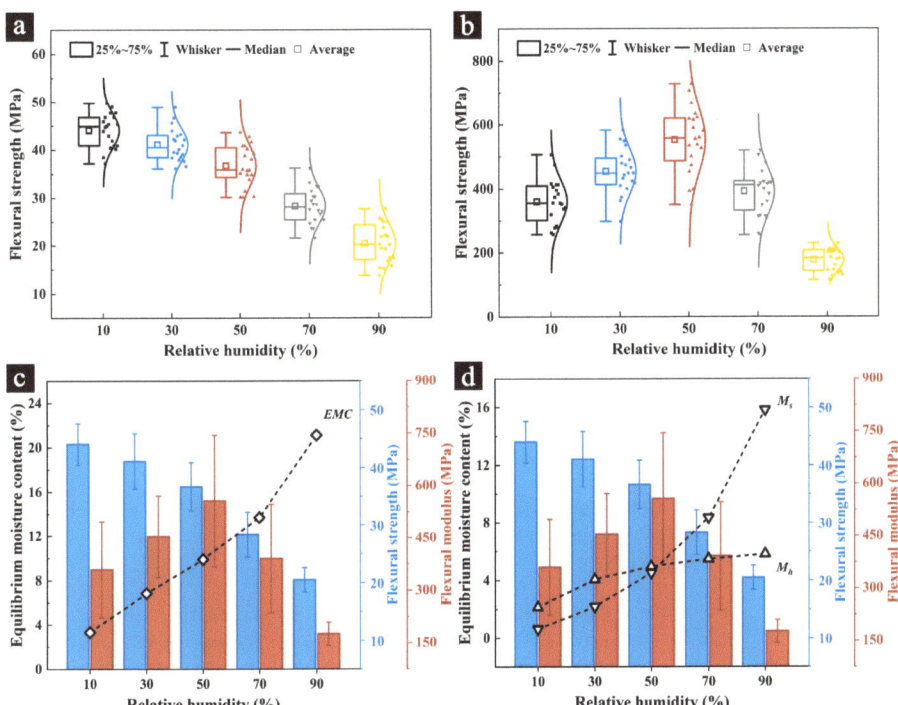

Figure 3. The flexural strength (**a**) and flexural modulus (**b**) of the samples at different humidity levels during the desorption process, along with the correlations between the flexural strength and modulus with the EMC (**c**), and M_h and M_s (**d**).

However, different from this, the flexural strength and modulus during the desorption process were generally lower than during the adsorption process at the same relative humidity. For example, at 10% RH, the flexural strength and modulus during the adsorption process were 44.86 MPa and 380.38 MPa, respectively, while during the desorption process, they were 44.04 MPa and 360.06 MPa. At 90% RH, the flexural strength and modulus during adsorption were 22.03 MPa and 208.06 MPa, respectively, whereas during desorption, they dropped to 20.50 MPa and 176.83 MPa.

The isothermal moisture desorption curve of the samples (Figure 3c) and the variation of M_h and M_s with humidity (Figure 3d) further confirm the changes in the mechanical properties of the samples. Due to the hysteresis phenomenon, the EMC during the desorption process was consistently higher than during the adsorption process. The EMC values of the samples at 10% RH, 30% RH, 50% RH, 70% RH, and 90% RH during adsorption were 2.43%, 5.16%, 7.62%, 10.60%, and 18.85%, respectively, while during desorption, they were 3.33%, 6.85%, 9.85%, 13.67%, and 21.10%, respectively. This increase in moisture content led to a decrease in the mechanical strength and stiffness of the samples. Additionally, the results from the Hailwood–Horrobin (H–H) model suggest that this change is primarily driven by the increase in M_h, indicating that monolayer adsorption plays a significant role in the observed reduction in the mechanical properties.

2.2.3. Hysteresis

A comparison of the flexural strength and flexural modulus of the samples subject to different relative humidity conditions during the two processes (Figure 4a) and their relationship with desorption hysteresis (Figure 4b) shows clear differences. As mentioned earlier, both the flexural strength and modulus during the desorption process are significantly lower than those during the adsorption process at the same relative humidity, which is caused by the hysteresis effect on the samples. The extent of the reduction in the flexural strength or modulus varies according to different humidity conditions.

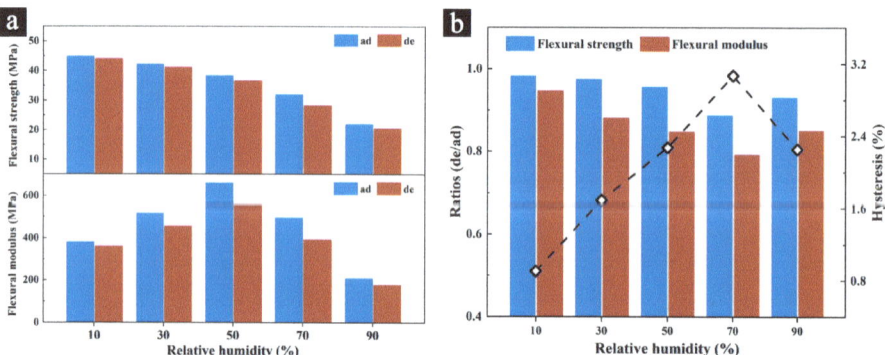

Figure 4. The changes in flexural strength and flexural modulus of the samples for different humidity levels during the two processes (**a**), and the correlation between these changes and the hysteresis of the samples (**b**).

The ratio of the flexural strength between the desorption and adsorption processes (de/ad) decreases as the humidity increases, with values of 0.98, 0.97, 0.95, 0.88, and 0.93, respectively, from low to high humidity. Similarly, the ratio of the flexural modulus (de/ad) follows the trend of 0.95, 0.88, 0.85, 0.79, and 0.85, showing a pattern of first decreasing and then increasing with increasing humidity. Both ratios reach their lowest point at 70% RH, indicating that the changes in the flexural strength and modulus are most pronounced at this humidity level.

Hysteresis typically represents the hygroscopic stability and adsorption efficiency of a material. The larger the hysteresis value, the slower the material absorbs or releases moisture when humidity fluctuates, and the more moisture it retains [36]. The ratio of the flexural strength and flexural modulus between the two processes is inversely proportional to the hysteresis value. In other words, the larger the hysteresis value at a particular humidity level, the greater the changes in the mechanical properties of the sample. This indicates that not only do the mechanical properties of the sample change with humidity fluctuations, but these changes are further amplified by the effect of hysteresis. If the sample is exposed to an environment with frequent humidity fluctuations over an extended period, continuous adsorption and desorption processes can lead to irreversible collapse of the pore structure in the samples, which, at the macroscopic level, results in reduced flexibility and decreased mechanical strength, thereby further impacting the overall mechanical performance of the material. Particularly in common indoor conditions ranging from 50% RH to 70% RH, palm leaf manuscripts may not only suffer from external mechanical damage, but could also experience internal stress due to changes in their mechanical properties. This internal stress may lead to issues, such as cracking and other forms of deterioration [28].

The results of the short-term study on the effects of relative humidity on the mechanical properties of the samples indicate that different humidity levels lead to changes in the EMC of palm leaf manuscripts, which in turn significantly affects their mechanical properties. Maintaining a stable environment in terms of humidity can prevent changes in the mechanical properties of palm leaf manuscripts, thereby preventing deterioration. Additionally, the samples generally exhibited better mechanical properties in relatively dry environments (≤50% RH), compared to higher humidity conditions (70% RH and 90% RH). This suggests that preserving palm leaf manuscripts in a drier environment would be a more ideal choice for their long-term conservation.

2.3. Long-Term Aging Effects of Relative Humidity on the Mechanical Properties of the Samples

2.3.1. Aging Results of the Samples

Figure 5 shows the microscopic morphological changes to the samples after 100 days of aging in different conditions (temperature of 25 °C and humidity levels of 10% RH, 30% RH, 50% RH, 70% RH, and 90% RH).

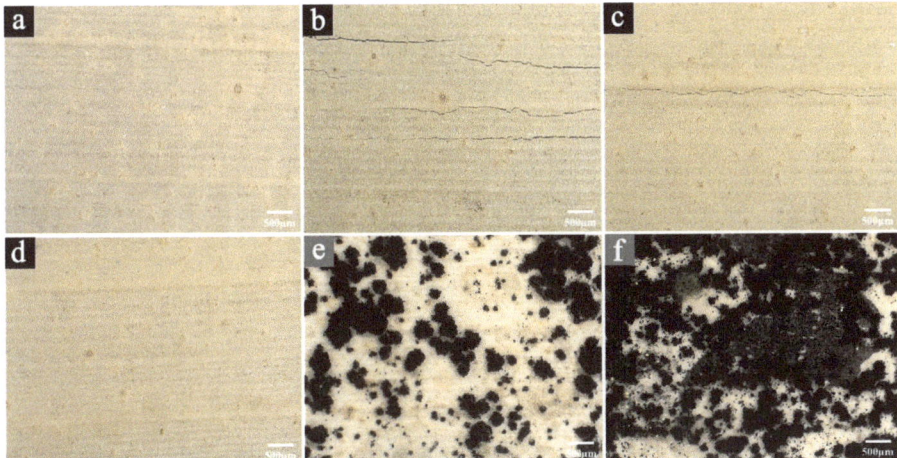

Figure 5. The microscopic images of the samples before aging (**a**) and after 100 days of aging in different humidity conditions: 10% RH (**b**), 30% RH (**c**), 50% RH (**d**), 70% RH (**e**), and 90% RH (**f**).

After 100 days of aging at 10% RH (Figure 5b) and 30% RH (Figure 5c), cracks appeared on the surface of the samples, with more severe cracking observed in the samples aged at 10% RH. In contrast, after 100 days of aging at 70% RH (Figure 5e) and 90% RH (Figure 5f), a significant amount of mold growth was observed on the surface of the samples, with the samples aged at 90% RH exhibiting more and denser mold colonies. Only the samples aged at 50% RH (Figure 5d) for 100 days showed no cracks or microbial damage, and their morphology did not significantly change compared to before aging.

2.3.2. Mechanical Properties

The mechanical properties of the samples after 100 days of aging in different conditions were measured using the TMA, represented by the flexural strength (Figure 6a) and flexural modulus (Figure 6b). The results indicate that the TMA is a viable method for measuring the mechanical properties of precious and relatively fragile organic materials, such as palm leaf manuscripts.

After 100 days of aging, the flexural strength of the samples in different aging conditions (Figure 6a) decreased to different degrees. The flexural strength of the samples aged at 10% RH and 30% RH for 100 days was 26.46 MPa and 28.71 MPa, respectively, which may be due to the cracking phenomenon on the surface of the samples. After 100 days of aging at 70% RH and 90% RH, the flexural strength of the samples decreased significantly, reaching only 8.11 MPa and 5.51 MPa, respectively, which may be due to the degradation of the cellulose and hemicellulose components in the samples due to the growth of mold on the surface of the samples and their propagation activities [27]. The sample, after aging in 50%

RH for 100 days, had the highest flexural strength (33.54 MPa) among all the aged samples, indicating that 50% RH had the least effect on the mechanical strength of the samples. After 100 days of aging, the change in the flexural modulus (Figure 6b) of the samples subject to different aging conditions was consistent with the change in the flexural strength and the sample, after 100 days of aging in 50% RH, had the highest flexural modulus (581.55 MPa) among all the aged samples.

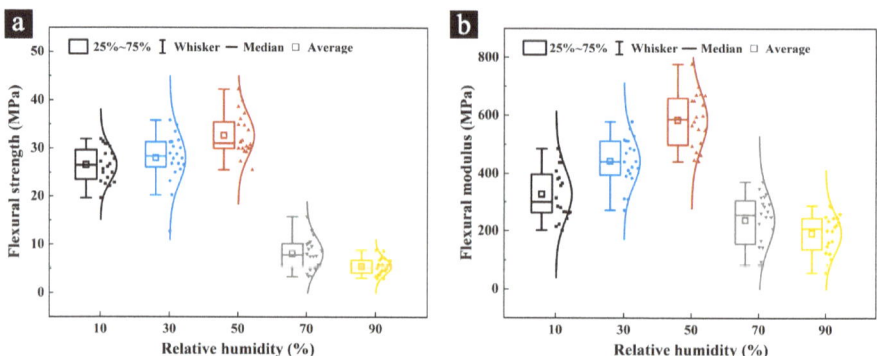

Figure 6. The flexural strength (**a**) and flexural modulus (**b**) of the samples after 100 days of aging in different humidity conditions.

2.3.3. FT-IR

The infrared spectra of the samples after aging for 100 days subject to different aging conditions are shown in Figure 7a. The relative peak intensities obtained from the semi-quantitative analysis of the IR spectra of the samples are shown in Figure 7b.

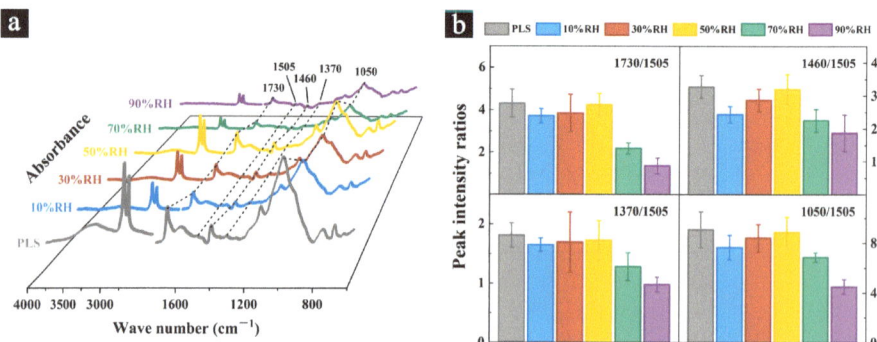

Figure 7. Infrared spectra of the samples before and after aging (**a**) and the ratio of characteristic peak intensities (**b**).

The results indicate that after 100 days of aging in different conditions, the relative intensities of the characteristic peaks in the infrared spectra of the samples for cellulose and hemicellulose showed varying degrees of reduction. In the samples aged at 70% RH for 100 days, the peak intensity ratios I_{1730}/I_{1505}, I_{1460}/I_{1505}, I_{1370}/I_{1505}, and I_{1050}/I_{1505} decreased by 49.32%, 31.21%, 29.04%, and 24.28%, respectively, compared to the values before aging. For samples aged at 90% RH for 100 days, these peak intensity ratios decreased by 68.62%, 42.62%, 45.61%, and 50.33%, respectively. This suggests that in high humidity environments, exacerbated by mold growth, the relative content of cellulose and hemicellulose in the samples decreases significantly, with greater degradation at higher humidity levels. Some studies have shown that the degradation of cellulose and hemicellulose can lead to a reduction in the mechanical properties of fibers, such as flexural strength and flexibility, potentially resulting in material damage related to physical performance [41]. Although the peak intensities of the samples aged at 10% RH and 30% RH for 100 days also decreased compared to the values before aging (with greater degradation observed in the 10% RH samples), the extent of the degradation was much less pronounced than in the high-humidity aged samples. This suggests that while dry environments do not significantly affect the relative content of cellulose and hemicellulose, prolonged exposure to excessively low humidity can cause irreversible issues, such as material shrinkage and surface cracking (as seen in Figure 5b,c), which also contribute to a decline in the material's mechanical properties. After 100 days of aging at 50% RH, the peak intensity values of the samples decreased by only 1.39%, 2.14%, 4.42%, and 1.99%, respectively, compared to the values before aging, making these the least degraded among all the aged samples.

2.3.4. XRD

The XRD patterns and the calculated relative crystallinity results of the samples after aging for 100 days subject to different aging conditions are shown in Figure 8a and 8b, respectively.

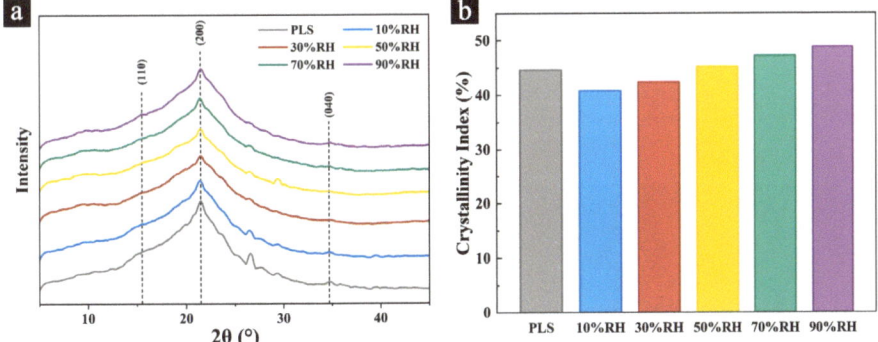

Figure 8. XRD patterns of the samples before and after aging (**a**) and the crystallinity index (**b**).

The results show that the (200) main peak around 22° and the characteristic peaks at the (110) crystal plane (around 15°) and the (040) crystal plane (around 34°) in all the samples did not exhibit significant shifts in position. This indicates that the crystalline structure of cellulose in all the samples is Type I and aging in different relative humidity levels did not cause noticeable changes to the interplanar spacing or structural rearrangement. However, the (200) diffraction peaks of the samples aged in high humidity conditions (70% RH and 90% RH) were sharper, compared to those aged in low humidity conditions, and the amorphous regions appeared to be shorter. This suggests that the crystallinity of cellulose in the samples may have undergone varying degrees of change. Additionally, some of the observed non-cellulose diffraction peaks (e.g., at 26° and 29°) may originate from crystalline substances, such as Ca and Si, introduced during the preparation of the palm leaf manuscripts [7]. The calculated results on the relative crystallinity indicate that the crystallinity of the cellulose in the samples aged at 10% RH and 30% RH decreased compared to those before aging, likely due to minor cellulose degradation during the aging process. In contrast, the relative crystallinity of the cellulose in the samples aged at 70% RH and 90% RH increased compared to the pre-aging state. This suggests that while high humidity and mold invasion lead to significant degradation of both cellulose and hemicellulose, the amorphous regions, such as hemicellulose, degrade at a faster rate and to a greater extent, resulting in an apparent increase in the relative crystallinity of cellulose. This observation is consistent with the phenomena observed in the infrared spectroscopy analysis of the samples. In comparison, the crystallinity of the cellulose in the samples aged at 50% RH showed minimal changes, indicating a more stable structure.

The results of the study on the long-term aging effects of relative humidity on the mechanical properties of the samples show that excessively dry environments lead to surface cracking in palm leaf manuscripts, while humid environments result in mold growth and the degradation of key chemical components. Although the causes differ, both environments ultimately result in significant changes to the mechanical properties of palm leaf manuscripts, indicating that neither environment is suitable for their preservation. In contrast, the samples aged at 50% RH exhibited no visible damage, and their mechanical properties and chemical structure remained largely unchanged. This suggests that 50% RH is a relatively optimal humidity condition for the preservation of palm leaf manuscripts.

3. Materials and Methods

3.1. Experimental Materials

The experimental samples in this study were raw palm leaf manuscripts from Yunnan Province, China. These samples were produced following the traditional palm-leaf manuscript-making process, which has been recognized as part of China's national intangible cultural heritage. The preparation involved boiling, washing, air drying, trimming, and flattening the leaves from the talipot palm (*Corypha umbraculifera* L.) tree [11], resulting in the creation of the samples used in this study. This research focuses on the impact of relative humidity on the mechanical properties of the palm leaves themselves, the core material of palm leaf manuscripts. Therefore, the raw palm leaf manuscript samples (PLSs) were not subjected to subsequent treatments, such as writing or coloring, in order to avoid any influence from pigments or binding materials on the study results. A photograph of a raw palm leaf manuscript sample is shown in Figure 9.

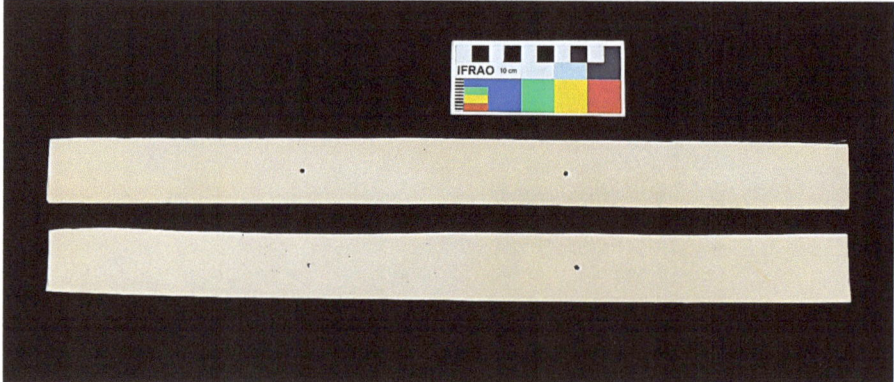

Figure 9. Photograph of a raw palm leaf manuscript sample (PLS).

3.2. Simultaneous DVS

The equilibrium moisture content (EMC) of the samples, as well as the isothermal moisture adsorption and desorption curves within the tested humidity range, were measured using a high-throughput dynamic vapor sorption analyzer (SPSx-1μ, ProUmid, Ulm, Germany) subject to the same temperature, but different relative humidity conditions. The testing range was 0% to 95% RH, with humidity increments of 10% RH between 0% and 90% RH, and 5% RH between 90% and 95% RH. The temperature was maintained at 25 °C. The equilibrium conditions were defined as a weight change of less than 0.1% within 10 min, or a maximum measurement time of 360 min, per humidity gradient.

3.3. Sorption Models

The hygroscopic test results were fitted using the classic Hailwood–Horrobin (H–H) model and the model parameters were calculated using software (SPSS Statistics 22, IBM, Armonk, NY, USA) to further explain the hygroscopic characteristics of the samples.

The H–H model equation is as follows:

$$\text{EMC} = M_h + M_s = \frac{1800}{w} \cdot \frac{k_1 \cdot k_2 \cdot \text{RH}}{100 + k_1 \cdot k_2 \cdot \text{RH}} \cdot 100\% + \frac{1800}{w} \cdot \frac{k_2 \cdot \text{RH}}{100 - k_2 \cdot \text{RH}} \quad (1)$$

where EMC (g/g) is the equilibrium moisture content; RH (%) is the relative humidity; M_h is the monolayer moisture content (%); M_s is the multilayer moisture content (%); w is the molecular weight of the materials at each adsorption site; and k_1 and k_2 are equilibrium constants in the sorption process [31].

3.4. Flexural Strength Test

The flexural strength of the samples was tested using a thermomechanical analyzer (TMA 7100, Hitachi, Chiyoda, Japan), equipped with quartz probes and components, as shown in Figure 10(a1). To investigate the short-term effects of relative humidity on the mechanical properties of palm leaf manuscripts, a custom-made humidity control chamber was connected to the TMA. The stable relative humidity environment was provided by a Modular Humidity Generator (MHG 32, ProUmid, Ulm, Germany), as illustrated in Figure 10(a2).

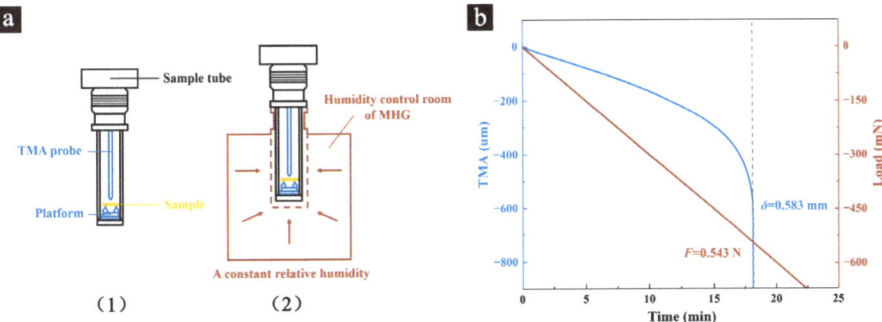

Figure 10. A schematic diagram of the TMA's flexural strength test components (**a1**) and the humidity application process (**a2**), along with the load/displacement–time curves during the testing process (**b**).

Two relative humidity control programs were used to simulate the changes in the samples during the adsorption and desorption phases, as follows:

- Adsorption phase: After controlling the relative humidity at 0% RH for 6 h, the humidity was increased to the set levels (10% RH, 30% RH, 50% RH, 70% RH, and 90% RH) and held for 6 h;
- Desorption phase: After controlling the relative humidity at 95% RH for 6 h, the humidity was decreased to the set levels (same as above) and held for 6 h.

After completing these phases, the flexural strength test was conducted at a constant temperature of 25 °C, with the tested humidity set to the corresponding levels. The initial load was 0.1 mN and the loading rate was 30 mN/min, continuing until the sample fractured and the test curves were obtained (Figure 2b). The samples were cut to approximately 8 mm × 1 mm, with 20 samples tested under each humidity condition.

For the aged samples, no humidity was applied during testing. The TMA was used to test the samples directly. Similarly, the samples were cut to approximately 8 mm × 1 mm, with 20 samples tested under each aging condition.

The flexural strength (σ) and flexural modulus (E) of the samples were calculated using Formulas (2) and (3), respectively:

$$\sigma = \frac{3FL}{2bd^2} \quad (2)$$

$$E = \frac{FL^3}{4\delta bd^3} \tag{3}$$

where σ (MPa) and E (MPa) are the flexural strength and flexural modulus of the sample, respectively; F (N) is the maximum load at the point of sample fracture; L (mm) is the span between the supports on the TMA's flexural strength testing platform, fixed at 5 mm; b (mm) and d (mm) are the width and thickness of the sample, respectively, measured using a super-depth microscope (VHX-6000, KEYENCE, Osaka, Japan); δ (mm) is the maximum displacement of the sample at the point of fracture.

3.5. Simulated Aging Experiment

A sufficient number of raw palm leaf manuscript samples (PLSs) were divided into five groups, each containing 30 individual samples. The five groups of samples were placed in an environmental test chamber (GSH-64, Espec, Osaka, Japan), where different aging conditions were set (all at a temperature of 25 °C, with humidity levels of 10% RH, 30% RH, 50% RH, 70% RH, and 90% RH). After 100 days of aging, the samples were removed and their morphological changes were observed using a super-depth microscope (VHX-6000, KEYENCE, Osaka, Japan). Some of the samples were then selected for flexural strength testing and chemical structure analysis.

3.6. FT-IR Test

The aged samples were finely ground and passed through a 200-mesh sieve. They were then mixed with KBr (spectrally pure, Macklin, Shanghai, China) at a mass ratio of 1:100 and pressed into pellets. The chemical structure of the samples was characterized using Fourier transform infrared spectroscopy (Nicolet™ iS™5, Thermo Scientific, Waltham, MA, USA), to compare the differences in the characteristic functional groups before and after aging. The absorption spectra were recorded in the range of 4000–800 cm^{-1}, with baseline corrections performed at 4000, 1890, 1530, and 864 cm^{-1} [42]. Three samples were tested for each aging condition.

In order to further characterize the changes to the chemical structure of the samples, the infrared spectra of the samples were analyzed semi-quantitatively, using the peak intensity method. The absorption peaks at 1730 cm^{-1} (stretching vibration by the carbonyl group on hemicellulose) and 1460 cm^{-1} (stretching vibration by the methylene group on hemicellulose) were selected as the characteristic peaks of hemicellulose, and the absorption peak at 1505 cm^{-1} (backbone vibration of butyl propane in butyl lignin) was chosen to represent lignin, and the absorption peaks at 1370 cm^{-1} (bending vibration by the methyl group on cellulose) and 1050 cm^{-1} (bending vibration by the glycosidic bond on cellulose) were chosen to represent cellulose [43–45]. The intensity values of the infrared spectral peaks were measured after baseline correction using software (OMNIC 9.2, Thermo Scientific, USA).

3.7. XRD Test

The crystal structure and crystallinity of the aged samples were analyzed using an X-ray diffractometer (D8 ADVANCE, Bruker, Munich, Germany). The operational settings included a working voltage of 40 kV, a current of 40 mA, and the use of Cu-Kα radiation (wavelength λ = 1.5406 Å), as the radiation source. The scanning angle range during measurement was set from 5° to 50° (2θ), with a scanning speed of 1°/min and a step size of 0.02°. The diffraction patterns were fitted using software (MDI Jade 9, ICDD, Newtown Square, PA, USA) and the crystallinity index (CI) of the samples was calculated using the Segal method.

$$CI = \frac{I_{200} - I_{am}}{I_{200}} \times 100\% \tag{4}$$

where CI is the crystallinity index of the sample; I_{200} is the maximum intensity of the lattice diffraction angle of (200) near 2θ = 22.4°, which signifies both the crystalline and non-crystalline regions; and I_{am} is the minimum intensity near the 2θ angle of 18°, indicating the non-crystalline region.

4. Conclusions

This study shows that varying levels of relative humidity affect the EMC and significantly impact the mechanical properties of palm leaf manuscripts. The flexural strength decreased notably with higher humidity, while the flexural modulus initially increased and then decreased as the humidity rose. Hysteresis caused a reduction in both the flexural strength and modulus during desorption compared to adsorption, with larger hysteresis values resulting in greater changes to the mechanical properties. Therefore, maintaining a stable environment in regard to humidity can help preserve the mechanical integrity of palm leaf manuscripts and prevent deterioration.

Long-term exposure to either very dry or humid conditions harms the mechanical properties of palm leaf manuscripts, making both conditions unsuitable for preservation. Dry conditions lead to surface cracking, while humid conditions promote mold growth, which degrades the manuscripts' primary chemical components. Samples aged at 50% RH showed no visible damage, with the material's mechanical properties and chemical structure largely preserved, suggesting that 50% RH is an optimal humidity condition for preserving palm leaf manuscripts.

This study used DVS and TMA-MHG methods to examine changes in the mechanical properties of palm leaf manuscripts subject to various humidity conditions, assessing both the short-term effects and long-term aging. The results provide a comprehensive and in-depth understanding of the characteristics of palm leaf manuscripts, aiding in the development of more effective preventive conservation strategies. It is important to note, however, that the preservation of palm leaf manuscripts is often influenced by a combination of temperature and humidity. The recommendation of 50% RH as a suitable humidity condition is based on preliminary data from this study. Future research should explore the combined effects of more complex environmental factors on both the mechanical and other properties of palm leaf manuscripts. In summary, this study provides valuable data that support the long-term preservation and preventive conservation of these priceless manuscripts, providing a useful reference for further research.

Author Contributions: Conceptualization, W.Z., S.W. and H.G.; methodology, W.Z.; software, W.Z.; validation, S.W. and H.G.; formal analysis, H.G.; investigation, W.Z.; resources, H.G.; data curation, W.Z.; writing—original draft preparation, W.Z.; writing—review and editing, S.W. and H.G.; visualization, W.Z.; supervision, S.W.; project administration, H.G.; funding acquisition, H.G. All authors have read and agreed to the published version of the manuscript.

Funding: This research was funded by the National Key Research and Development Program of China, grant number 2022YFF0903905.

Institutional Review Board Statement: Not applicable.

Informed Consent Statement: Not applicable.

Data Availability Statement: The data presented in this study are available on request from the corresponding author.

Acknowledgments: The authors express their gratitude to Liusan Li, Li Li, and Liuyang Han for their assistance in this study.

Conflicts of Interest: The authors declare that there are no conflicts of interest.

References

1. Kumar, D.U.; Sreekumar, G.; Athvankar, U. Traditional writing system in southern India—Palm leaf manuscripts. *Des. Thoughts* **2009**, *7*, 2–7.
2. Panigrahi, A.K.; Litt, D. Odia Script in Palm-Leaf Manuscripts. *J. Humanit. Soc. Sci.* **2018**, *23*, 13–19.
3. Meher, R. Tradition of palm leaf manuscripts in Orissa. *Orissa Rev.* **2009**, 43–46.

4. Alexander, T.J.; Kumar, S.S. A novel binarization technique based on Whale Optimization Algorithm for better restoration of palm leaf manuscript. *J. Ambient. Intell. Humaniz. Comput.* **2020**, 1–8. [CrossRef]
5. Subramani, K.; Subramaniam, M. Creation of original Tamil character dataset through segregation of ancient palm leaf manuscripts in medicine. *Expert Syst.* **2021**, *38*, e12538. [CrossRef]
6. Sabeenian, R.; Paramasivam, M.; Anand, R.; Dinesh, P. (Eds.) Palm-leaf manuscript character recognition and classification using convolutional neural networks. In *Computing and Network Sustainability: Proceedings of IRSCNS 2018, Panaji, India, 30–31 August 2018*; Springer: Berlin/Heidelberg, Germany, 2019. [CrossRef]
7. Chu, S.; Lin, L.; Tian, X. Evaluation of the Deterioration State of Historical Palm Leaf Manuscripts from Burma. *Forests* **2023**, *14*, 1775. [CrossRef]
8. Sharma, D.; Singh, M.; Krist, G.; Velayudhan, N.M. Structural characterisation of 18th century Indian Palm leaf manuscripts of India. *Int. J. Conserv. Sci.* **2018**, *9*, 257–264.
9. Sharma, D.; Singh, M.R.; Dighe, B. Chromatographic Study on Traditional Natural Preservatives Used for Palm Leaf Manu-scripts in India. *Restaurator* **2018**, *39*, 249–264. [CrossRef]
10. Singh, M.R.; Sharma, D. Investigation of Pigments on an Indian Palm Leaf Manuscript (18th–19th century) by SEM-EDX and other Techniques. *Restaurator* **2020**, *41*, 49–65. [CrossRef]
11. Agrawal, O.P. *Conservation of Manuscripts and Paintings of South-East Asia*; Butterworth-Heinemann: London, UK, 1984.
12. Sah, A. Palm Leaf Manuscripts of the World: Material, Technology and Conservation. *Stud. Conserv.* **2002**, *47* (Suppl. S1), 15–24. [CrossRef]
13. Wiland, J.; Brown, R.; Fuller, L.; Havelock, L.; Johnson, J.; Kenn, D.; Kralka, P.; Muzart, M.; Pollard, J.; Snowdon, J. A literature review of palm leaf manuscript conservation—Part 2: Historic and current conservation treatments, boxing and storage, reli-gious and ethical issues, recommendations for best practice. *J. Inst. Conserv.* **2023**, *46*, 64–91. [CrossRef]
14. Wiland, J.; Brown, R.; Fuller, L.; Havelock, L.; Johnson, J.; Kenn, D.; Kralka, P.; Muzart, M.; Pollard, J.; Snowdon, J. A literature review of palm leaf manuscript conservation—Part 1: A historic overview, leaf preparation, materials and media, palm leaf manuscripts at the British Library and the common types of damage. *J. Inst. Conserv.* **2022**, *45*, 236–259. [CrossRef]
15. Zhang, M.; Song, X.; Wang, J.; Lyu, X. Preservation characteristics and restoration core technology of palm leaf manuscripts in Potala Palace. *Arch. Sci.* **2022**, *22*, 501–519. [CrossRef]
16. Joshi, Y. Modern techniques of preservation and conservation of palm leaf manuscripts. In Proceedings of the Conference on Palm Leaf and Other Manuscripts in Indian Languages, Puducherry, India, 11–13 January 1995.
17. Suryawanshi, D.G.; Nair, M.V.; Sinha, P.M. Improving the Flexibility of Palm Leaf. *Restaur. Int. J. Preserv. Libr. Arch. Mater.* **1992**, *13*, 37–46. [CrossRef]
18. Zhang, M.; Song, X.; Wang, Y. Two Different Storage Environments for Palm Leaf Manuscripts: Comparison of Deterioration Phenomena. *Restaur. Int. J. Preserv. Libr. Arch. Mater.* **2021**, *42*, 147–168. [CrossRef]
19. Zhang, W.; Wang, S.; Guo, H. Study on the Effects of Temperature and Relative Humidity on the Hygroscopic Properties of Palm Leaf Manuscripts. *Forests* **2024**, *15*, 1816. [CrossRef]
20. Graminski, E.L.; Parks, E.J.; Toth, E.E. The Effects of Temperature and Moisture on the Accelerated Aging of Paper. *Restaurator* **1979**, *3*, 175–199.
21. Liu, X.Y.; Timar, M.C.; Varodi, A.M.; Sawyer, G. An Investigation of Accelerated Temperature-Induced Ageing of Four Wood Species: Colour and FTIR. *Wood Sci. Technol.* **2017**, *51*, 357–378. [CrossRef]
22. Botti, S.; Di Lazzaro, P.; Flora, F.; Mezi, L.; Murra, D. Raman spectral mapping reveal molecular changes in cellulose aging induced by ultraviolet and extreme ultraviolet radiation. *Cellulose* **2024**, *31*, 749–758. [CrossRef]
23. Mitsui, K.; Takada, H.; Sugiyama, M.; Hasegawa, R. Changes in the Properties of Light-Irradiated Wood with Heat Treatment: Part 1. Effect of Treatment Conditions on the Change in Color. *Holzforschung* **2001**, *55*, 601–605. [CrossRef]
24. Havermans, J. Effects of Air Pollutants on the Accelerated Ageing of Cellulose-Based Materials. *Restaurator* **1995**, *16*, 209–233. [CrossRef]
25. Bartl, B.; Mašková, L.; Paulusová, H.; Smolík, J.; Bartlová, L.; Vodicka, P. The Effect of Dust Particles on Cellulose Degradation. *Stud. Conserv.* **2016**, *61*, 203–208. [CrossRef]
26. Bylund Melin, C.; Hagentoft, C.-E.; Holl, K.; Nik, V.M.; Kilian, R. Simulations of Moisture Gradients in Wood Subjected to Changes in Relative Humidity and Temperature Due to Climate Change. *Geosciences* **2018**, *8*, 378. [CrossRef]
27. Kim, M.-J.; Choi, Y.-S.; Oh, J.-J.; Kim, G.-H. Experimental investigation of the humidity effect on wood discoloration by se-lected mold and stain fungi for a proper conservation of wooden cultural heritages. *J. Wood Sci.* **2020**, *66*, 31. [CrossRef]
28. Yu, D.; Li, X.; Sun, S.; Guo, H.; Luo, H.; Zhu, J.; Li, L.; Wang, S.; Han, L. The effect of traditional processing craft on the hy-groscopicity of palm leaf manuscripts. *Herit. Sci.* **2024**, *12*, 280. [CrossRef]
29. Ziegler, J. Testing Aquazol®. An Evaluation of its Suitability for the Conservation of Works on Paper. *J. Pap. Conserv.* **2022**, *23*, 48–58. [CrossRef]
30. Liu, X.; Tu, X.; Ma, W.; Zhang, C.; Huang, H.; Varodi, A.M. Consolidation and Dehydration of Waterlogged Archaeological Wood from Site Huaguangjiao No.1. *Forests* **2022**, *13*, 1919. [CrossRef]
31. Han, L.; Yu, D.; Liu, T.; Han, X.; Xi, G.; Guo, H. Size Effect on Hygroscopicity of Waterlogged Archaeological Wood by Sim-ultaneous Dynamic Vapour Sorption. *Forests* **2023**, *14*, 519. [CrossRef]

32. Wu, M.; Mu, L.; Zhang, Z.; Han, X.; Guo, H.; Han, L. Anti-Cracking TEOS-Based Hybrid Materials as Reinforcement Agents for Paper Relics. *Molecules* **2024**, *29*, 1834. [CrossRef] [PubMed]
33. Wu, M.; Han, X.; Qin, Z.; Zhang, Z.; Xi, G.; Han, L. A Quasi-Nondestructive Evaluation Method for Physical-Mechanical Properties of Fragile Archaeological Wood with TMA: A Case Study of an 800-Year-Old Shipwreck. *Forests* **2022**, *13*, 38. [CrossRef]
34. Jamali, A.; Kouhila, M.; Ait Mohamed, L.; Jaouhari, J.T.; Idlimam, A.; Abdenouri, N. Sorption isotherms of Chenopodium ambrosioides leaves at three temperatures. *J. Food Eng.* **2006**, *72*, 77–84. [CrossRef]
35. Thommes, M.; Kaneko, K.; Neimark, A.V.; Olivier, J.P.; Rodriguez-Reinoso, F.; Rouquerol, J.; Sing, K.S.W. Physisorption of gases, with special reference to the evaluation of surface area and pore size distribution (IUPAC Technical Report). *Pure Appl. Chem.* **2015**, *87*, 1051–1069. [CrossRef]
36. Chen, Q.; Wang, G.; Ma, X.-X.; Chen, M.-L.; Fang, C.-H.; Fei, B.-H. The effect of graded fibrous structure of bamboo (*Phyllostachys edulis*) on its water vapor sorption isotherms. *Ind. Crops Prod.* **2020**, *151*, 112467. [CrossRef]
37. Auernhammer, J.; Keil, T.; Lin, B.; Schäfer, J.-L.; Xu, B.-X.; Biesalski, M.; Stark, R.W. Mapping Humidity-Dependent Mechanical Properties of a Single Cellulose Fibre. *Cellulose* **2021**, *28*, 8313–8332. [CrossRef]
38. Sala, C.M.; Robles, E.; Gumowska, A.; Wronka, A.; Kowaluk, G. Influence of Moisture Content on the Mechanical Properties of Selected Wood-Based Composites. *BioResources* **2020**, *15*, 5503–5513. [CrossRef]
39. Ham, C.-H.; Youn, H.J.; Lee, H.L. Influence of Fiber Composition and Drying Conditions on the Bending Stiffness of Paper. *BioResources* **2020**, *15*, 9197–9211. [CrossRef]
40. Patera, A.; Derluyn, H.; Derome, D.; Carmeliet, J. Influence of sorption hysteresis on moisture transport in wood. *Wood Sci. Technol.* **2016**, *50*, 259–283. [CrossRef]
41. Bergander, A.; Salmén, L. Cell wall properties and their effects on the mechanical properties of fibers. *J. Mater. Sci.* **2002**, *37*, 151–156. [CrossRef]
42. Guo, J.; Xiao, L.; Han, L.; Wu, H.; Yang, T.; Wu, S.; Yin, Y. Deterioration of the cell wall in waterlogged wooden archeological artifacts, 2400 years old. *IAWA J.* **2019**, *40*, 820–844. [CrossRef]
43. Schwanninger, M.; Rodrigues, J.C.; Pereira, H.; Hinterstoisser, B. Effects of short-time vibratory ball milling on the shape of FT-IR spectra of wood and cellulose. *Vib. Spectrosc.* **2004**, *36*, 23–40. [CrossRef]
44. Marchessault, R.H. Application of infra-red spectroscopy to cellulose and wood polysaccharides. *Pure Appl. Chem.* **1962**, *5*, 107–130. [CrossRef]
45. Pandey, K.K.; Pitman, A.J. FTIR studies of the changes in wood chemistry following decay by brown-rot and white-rot fungi. *Int. Biodeterior. Biodegrad.* **2003**, *52*, 151–160. [CrossRef]

Disclaimer/Publisher's Note: The statements, opinions and data contained in all publications are solely those of the individual author(s) and contributor(s) and not of MDPI and/or the editor(s). MDPI and/or the editor(s) disclaim responsibility for any injury to people or property resulting from any ideas, methods, instructions or products referred to in the content.

Article

Thermal, Rheological, Structural and Adhesive Properties of Wheat Starch Gels with Different Potassium Alum Contents

Haibo Zhao [1], Hongbin Zhang [1], Qiang Xu [1], Hongdong Zhang [2,*] and Yuliang Yang [1,2,*]

[1] Institute for Preservation and Conservation of Chinese Ancient Books, Fudan University Library, Fudan University, Shanghai 200433, China
[2] State Key Laboratory of Molecular Engineering of Polymers, Department of Macromolecular Science, Fudan University, Shanghai 200433, China
* Correspondence: zhanghd@fudan.edu.cn (H.Z.); yuliangyang@fudan.edu.cn (Y.Y.)

Abstract: Wheat starch (WS) is a common adhesive material used in mounting of calligraphy and paintings. Potassium alum (PA) has indeed been used for many centuries to modify the physicochemical properties of starch. Thermal analysis revealed that the presence of PA led to an increase in the gelatinization temperature and enthalpy of the starch gels. The leached amylose and the swelling power of the starch gels exhibited a maximum at the ratio of 100:6.0 (WS:PA, w/w). The rheological properties of starch gels were consistent with changes in the swelling power of starch granules. SEM observations confirmed that the gel structure became more regular, and the holes grew larger with the addition of PA below the ratio of 100:6.0 (WS:PA, w/w). The short-range molecular order in the starch gels was enhanced by the addition of PA, confirmed by FT-IR analysis. Mechanical experiments demonstrated that the binding strength of the starch gels increased with higher PA concentrations and decreased significantly after the aging process. TGA results revealed that PA promoted the acid degradation of starch molecules. This study provides a detailed guide for the preparation of starch-based adhesive and its applications in paper conservation.

Keywords: wheat starch; potassium alum; adhesive

Citation: Zhao, H.; Zhang, H.; Xu, Q.; Zhang, H.; Yang, Y. Thermal, Rheological, Structural and Adhesive Properties of Wheat Starch Gels with Different Potassium Alum Contents. *Molecules* 2023, 28, 6670. https://doi.org/10.3390/molecules28186670

Academic Editor: Maria Luisa Saladino

Received: 21 August 2023
Revised: 4 September 2023
Accepted: 13 September 2023
Published: 17 September 2023

Copyright: © 2023 by the authors. Licensee MDPI, Basel, Switzerland. This article is an open access article distributed under the terms and conditions of the Creative Commons Attribution (CC BY) license (https://creativecommons.org/licenses/by/4.0/).

1. Introduction

Mounting of calligraphy and paintings is an ancient Chinese art form that has been practiced for over a millennium, serving as a means to conserve and preserve these exquisite works. During this process, a crucial element employed as an adhesive is a meticulously prepared starch gel derived from wheat starch [1]. Wheat starch is obtained by carefully washing the dough of wheat flour to remove gluten [2]. After the gelatinization process of starch, starch gels have the ability to form hydrogen bonds with cellulose fibers present in artworks [3]. This bonding mechanism is crucial in ensuring the longevity and stability of the artwork. However, the naturally derived starch gel has certain limitations, including relatively low viscosity and easy retrogradation [4,5]. Researchers have extensively investigated chemical, physical, and enzymatic modifications as means to overcome challenges associated with retrogradation and enhance the physicochemical properties of starch gels [6]. Among these approaches, physical modification techniques have gained considerable popularity due to their cost-effectiveness and minimal environmental impact [7]. Chinese restorers have ingeniously discovered that blending wheat starch with potassium alum (PA, KAl(SO$_4$)$_2$·12H$_2$O) yields a significant augmentation in the apparent viscosity and binding strength of starch gels as adhesive [8,9]. By embracing these innovative practices, the art of mounting calligraphy and paintings continues to thrive.

PA has already proven its significant effects on the properties of starch-based gels [10,11]. For instance, Li et al. demonstrated that PA could improve the tensile properties of potato starch dough and reduced breakage of the starch hydrogels [12]. Another study by Li et al. found that PA enhanced the stability of potato starch molecules, inhibited gelatinization, prevented starch molecule leakage, increased bound water content, and significantly enhanced the storage modulus [13]. Furthermore, Wang et al. reported that the effects of salts on the pasting and retrogradation of maize starch was in accord with the Hofmeister series [14]. In detail, potassium (K^+) and sulfate (SO_4^{2-}), as salting-out ions, were found to lower the swelling power and solubility of starch. In contrast, salting-in ion could increase these properties [15–18]. Moreover, Margit et al. observed that divalent cations such as calcium (Ca^{2+}) and magnesium (Mg^{2+}) had a greater impact in reducing starch recrystallization rate compared to monovalent cations [19]. PA consists of three ions, including two salting-out ions (K^+ and SO_4^{2-}) and one multivalent cation (Al^{3+}). As a result, the influence of PA on starch mixtures encompasses the comprehensive effects of these three ions working together.

The use of PA blended with wheat starch suspension to enhance the apparent viscosity and adhesive strength of starch gels has been a practice for many centuries. The ratios of WS:PA (w/w) used in restoration practices depend on the size of the works being restored. In general, the ratios range from 100:2 to 100:20 (WS:PA, w/w). However, the impact of PA on the apparent viscosity of wheat starch gels and its relationship with the binding capacity of the gel have not been thoroughly investigated. Furthermore, PA is an acid salt that has the potential to cause acid degradation and color change in paper relics, which has raised concerns among restorers [20,21]. Therefore, we aim to study the effects of PA on the thermal, rheological, and structural properties of wheat starch gels. Additionally, the relationship between adhesive strength and apparent viscosity of the starch gel was carefully determined. To evaluate the overall effects of PA on relics, including mechanical properties and color difference, aging experiments were conducted on paper samples bonded using WS-PA gels with various ratios. Through this comprehensive research, we aim to gain a deeper understanding of how PA influences the physicochemical properties of what starch gels and the influence of WS-PA gels on the conservation and preservation of precious paper relics.

2. Results
2.1. Thermal Properties

The gelatinization curves of the WS and WS-PA mixtures were analyzed by DSC, and the pasting parameters were calculated using specialized software. Figure 1 illustrates the gelatinization process, while Table 1 presents the values of the pasting parameters. As depicted in Figure 1, the addition of PA in the starch mixtures can delay the pasting process, and the extent of this delay is influenced by the concentration of PA. The pasting temperature always increases with the addition of PA. These temperature changes are related to the transition of the ordered structure in starch granules to amorphous structure during gelatinization [22]. Two reasons can explain this phenomenon. Firstly, the ions in PA can interact with water molecules and decrease the amount of free water available relative to starch, thereby enhancing the stability of the starch granules [13,23]. Secondly, the Al^{3+} ions present in PA can combine with water to form positively charged hydrated forms such as $Al(OH)^{2+}$, $Al_2(OH)_2^{4+}$, $Al_3(OH)_4^{5+}$, and $Al_{13}O_4(OH)_{24}^{7+}$ [24]. These hydrated products may act as crosslinking points, interacting with starch molecules in the amorphous regions or on the surface of starch granules. This crosslinking effect inhibits granule expansion and delays its destruction [14]. The gelatinization enthalpy (ΔH) is a measure of the energy required for the gelatinization process of starch and primarily characterizes the loss of the double helix structure and the melting of the crystalline structure within the starch [25]. The presence of PA significantly increased the gelatinization enthalpy of the starch mixtures. For instance, the enthalpy of a sample with ratio of 100:120 (WS:PA, w/w) increases by 8.7% compared to the blank sample. This increase indicates that more energy is needed to disrupt the starch granules. These findings are consistent with results concluded by Chen et al., who reported that multivalent cations can enhance the stability of the granules [12].

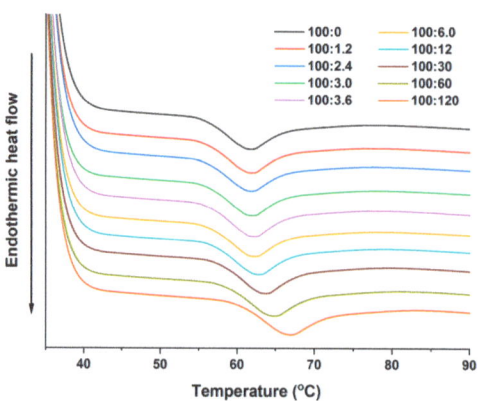

Figure 1. Gelatinization thermograms of WS and WS–PA mixtures.

Table 1. Gelatinization temperatures and enthalpy of WS and WS–PA mixtures.

WS:PA (w/w)	To (°C)	Tp (°C)	Tc (°C)	ΔH (J/g)
100:0	55.98 ± 0.17 a	61.30 ± 0.21 a	65.91 ± 0.33 a	1.525 ± 0.025 a
100:1.2	56.08 ± 0.02 ab	61.38 ± 0.05 ab	66.00 ± 0.24 a	1.542 ± 0.005 ab
100:2.4	56.20 ± 0.05 bc	61.53 ± 0.10 ab	66.08 ± 0.19 ab	1.548 ± 0.024 ab
100:3.0	56.23 ± 0.03 bc	61.59 ± 0.05 ab	66.18 ± 0.20 ab	1.558 ± 0.073 ab
100:3.6	56.31 ± 0.10 cd	61.65 ± 0.07 b	66.27 ± 0.06 ab	1.571 ± 0.003 ab
100:6.0	56.45 ± 0.13 d	61.66 ± 0.10 b	66.42 ± 0.18 b	1.578 ± 0.064 ab
100:12	56.96 ± 0.07 e	62.34 ± 0.12 c	66.94 ± 0.12 c	1.582 ± 0.015 ab
100:30	57.75 ± 0.07 f	63.21 ± 0.18 d	67.87 ± 0.04 d	1.602 ± 0.065 bc
100:60	58.68 ± 0.12 g	64.20 ± 0.27 e	69.04 ± 0.26 e	1.649 ± 0.082 c
100:120	60.17 ± 0.19 h	66.20 ± 0.37 f	71.21 ± 0.36 f	1.657 ± 0.046 c

The measurements were performed in triplicate. The mean ± SD values within the same column for each sample, followed by different lowercase letters, are significantly different ($p < 0.05$).

2.2. Leached Amylose and Swelling Power

The leached amylose of samples is an important parameter that indicates the leaching of soluble amylose starch and segmental amylopectin from starch granules during gelatinization. Swelling power represents the water absorption capacity of starch granules [26]. Figure 2 shows the leached amylose and swelling power of WS and WS–PA gels at various ratios. As shown in Figure 2, both leached amylose and swelling power of WS and WS–PA gels increase with higher temperature [27]. The higher temperature facilitates the disintegration of starch granules and the disruption of the crystalline structure, leading to the leaching of amylose and the breakage of the double helix structure of amylopectin. Figure 2a demonstrates that the leached amylose of WS–PA mixtures at 95 °C increases with the addition of PA below the ratio of 100:6.0 (WS:PA, w/w) and decreases with the addition of PA above the ratio of 100:6.0 (WS:PA, w/w). Based on Figure 2b, the swelling power of the blends at 95 °C also reaches its maximum at a ratio of 100:6.0 (WS:PA, w/w), which is consistent with the results of leached amylose measurements. Different types of ions can significantly affect the pasting behavior of starch, in accordance with the Hofmeister series [28]. PA consists of K^+, Al^{3+}, and SO_4^{2-} ions. Wang et al. discovered that SO_4^{2-} and K^+ ions, as salting-out agents, could improve the stability of starch granules [15]. Conversely, Al^{3+} ions, as multivalent cation, could bind with water molecules to form hydroxyl coordination compounds ($Al(OH)_x(H_2O)_y^n$) which may promote the disintegration of starch granules [24]. The results indicate that when the addition of PA is below the ratio of 100:6.0 (WS:PA, w/w), the PA could promote the disintegration of starch granules, with Al^{3+} ions playing a dominant role. However, when the addition of PA is above the ratio of 100:6.0 (WS:PA, w/w), the stability of starch granules is improved. There are two reasons to account for this phenomenon. Firstly, the SO_4^{2-} and K^+ ions play a dominant role as salt-out ions which could inhibit the breakage of starch granules. Secondly, the positively charged hydroxyl coordination compounds of Al^{3+} could act as cross-linking points through electrostatic interactions in the amorphous structure or on the surface of starch granules which inhibit the breakage of granules [14].

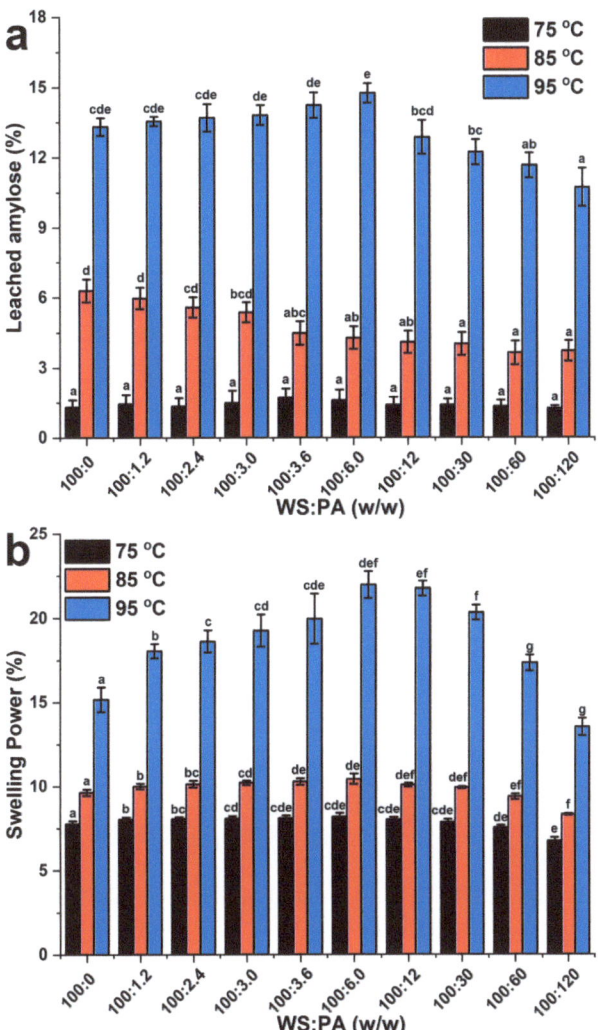

Figure 2. (a) The amount of leached amylose from WS and WS-PA gels at 75 °C, 85 °C and 95 °C, and (b) the swelling power of WS and WS-PA gels at 75 °C, 85 °C and 95 °C. The measurements were performed in triplicate. The mean ± SD values within the same column for each sample, followed by different lowercase letters, are significantly different ($p < 0.05$).

2.3. Rheological Measurements

2.3.1. Dynamic Rheological Measurements

The dynamic rheological properties of WS and WS–PA gels at different ratios of PA are analyzed and presented in Figure 3. The G′ and G″ of all samples increase with frequency, and the G′ is consistently higher than the G″ in all samples, indicating that the WS and WS–PA gels exhibit solid-like behavior [29]. In regards to the effect of PA concentration on the dynamic modulus, it is observed that when the addition of PA is below the ratio of 100:6.0 (WS:PA, w/w), the dynamic modulus increases with the presence of PA, as shown in Figure 3a. However, when the addition of PA is at higher concentrations, above the ratio of 100:6.0 (WS:PA, w/w), the dynamic modulus decreases as the PA concentration increases further. These findings are in agreement with the observed swelling power of the samples, suggesting that the WS–PA gel has the most compacted structure at the ratio of 100:6.0 (WS:PA, w/w). The results indicate that the breakage of granules reaches the maximum at this specific proportion and more-soluble starch leaching out from the granules participates in the network formation of WS–PA gels.

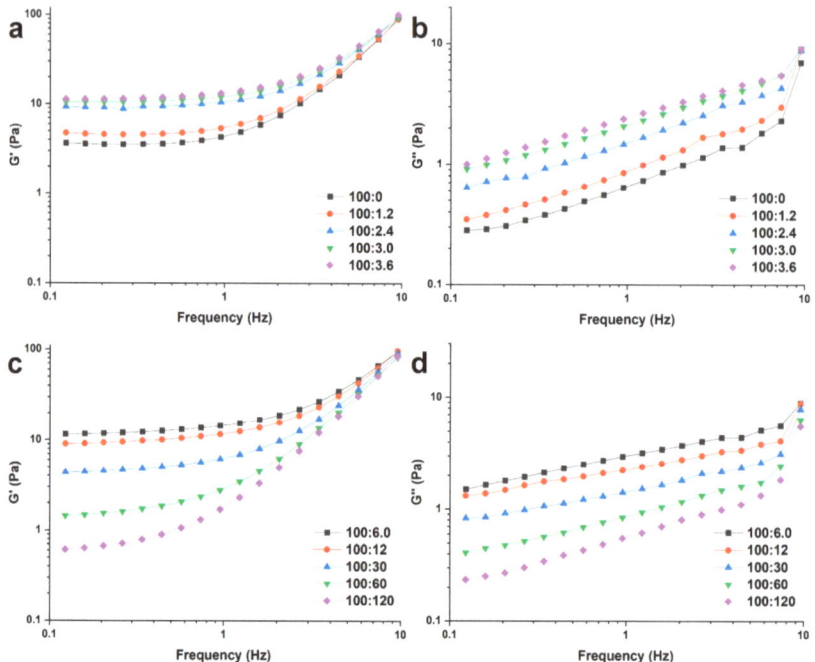

Figure 3. The dynamic rheological properties of WS and WS–PA gels. (**a**,**b**) Storage modulus (G′) and (**c**,**d**) loss modulus (G″) as a function of frequency at 25 °C and 1% strain.

2.3.2. Steady Rheological Measurements

Figure 4 presents the steady flow curves of WS and WS−PA gels, while Table 2 summarizes the fitting results of the steady flow curve data using the Herschel–Bulkley model. The R-squared values for all samples fitted by the Herschel–Bulkley model are above 0.964, indicating good model fit. All systems exhibit shear thinning behavior as the apparent viscosity of testing samples decreases with the increasing of shear rate. Additionally, the flow behavior index (n) values range from 0.51 to 0.77, which are significantly lower than 1. This suggests that both WS and WS−PA gels exhibit pseudoplastic fluid behavior with shear-thinning properties [30]. Moreover, it is observed that the apparent viscosity of WS−PA gels increases at low shear rates and reaches a maximum at a shear rate of 0.2 s^{-1}.

Firstly, this phenomenon can be attributed to the non-equilibration of the sample [31]. Secondly, when an external force is applied, the entanglement between the molecular chains of the material can gradually be disrupted, leading to relative displacement of the ordered structure under shear. This disruption of entanglements and the resulting relative displacement contribute to shear thickening. The addition of PA to WS gels has an impact on the apparent viscosity. The apparent viscosity increases with the augmentation ratios from 100:0 to 100:6.0 (WS:PA, w/w), but it then decreases as the ratios increase further from 100:6.0 to 100:120 (WS:PA, w/w). The consistency index (K) calculated by the Herschel–Bulkley model represents the viscosity of the gels. However, the behavior of K varied between ratios of 100:3.6, 100:6.0, and 100:12 (WS:PA, w/w). This decrease can be attributed to the destruction of the compact crosslinked structure of WS–PA gels at high shear rates, resulting in a significant decrease in viscosity and the lowest fitted value of K. The yield stress required to change the structure of the gels increases from 1.14 Pa to 7.74 Pa with the ratio from 100:0 to 100:6.0 (WS:PA, w/w). This indicates that more external force was required to alter the initial structure of starch gels at low addition concentrations of PA. However, the opposite trend is observed as the ratios increase from 100:6.0 to 100:120 (WS:PA, w/w) with the yield stress decreasing from 7.74 Pa to 0.37 Pa. This trend is consistent with the interpretation of the change in swelling power discussed in Section 2.2. Thixotropy is the ability of starch mixtures to recover their initial structure under external agitation or shear force [32]. The strength of thixotropy can be reflected by the area of the hysteretic loop, with a larger clockwise hysteretic loop indicating a more compact gel structure. The results indicate that the ratio at 100:6.0 (WS:PA, w/w) could result in obtaining the most compact structure of starch gels.

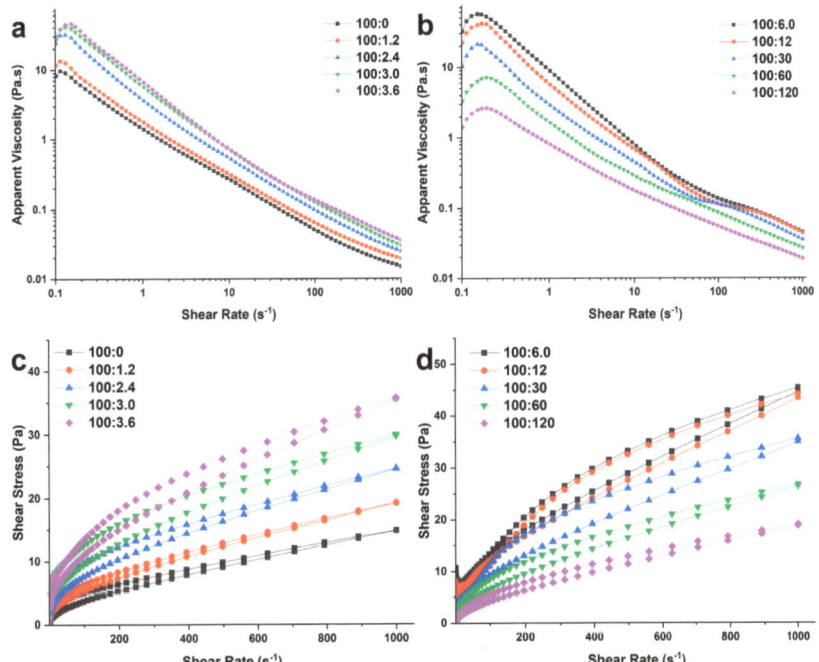

Figure 4. The steady rheological properties of WS and WS–PA gels. (**a**,**b**) Apparent viscosity and (**c**,**d**) shear stress as a function of shear rate.

Table 2. Steady flow parameters of WS and WS−PA gels.

WS:PA (w/w)	Up				Down				Hysteresis Loop
	τ_0 (Pa)	K (Pa·s^n)	n (-)	R^2	τ_0 (Pa)	K (Pa·s^n)	n (-)	R^2	
100:0	1.14 ± 0.24 ab	0.38 ± 0.12 bc	0.51 ± 0.04 a	0.992	0.55 ± 0.21 a	0.16 ± 0.08 a	0.65 ± 0.07 cd	0.998	668 ± 82 ab
100:1.2	1.48 ± 0.31 ab	0.41 ± 0.11 bcd	0.54 ± 0.03 ab	0.995	0.66 ± 0.09 a	0.31 ± 0.11 b	0.58 ± 0.03 ab	0.997	688 ± 97 a
100:2.4	3.34 ± 0.76 c	0.58 ± 0.04 def	0.51 ± 0.01 a	0.992	1.41 ± 0.05 b	0.49 ± 0.11 c	0.55 ± 0.02 a	0.998	1213 ± 483 b
100:3.0	4.97 ± 1.25 d	0.73 ± 0.11 e	0.51 ± 0.04 a	0.989	2.23 ± 0.45 cd	0.54 ± 0.11 c	0.56 ± 0.01 a	0.998	2069 ± 387 c
100:3.6	5.76 ± 0.44 d	0.48 ± 0.04 cde	0.61 ± 0.01 b	0.985	2.61 ± 0.22 de	0.57 ± 0.08 c	0.58 ± 0.01 ab	0.999	2160 ± 120 c
100:6.0	7.74 ± 1.34 e	0.21 ± 0.06 a	0.77 ± 0.05 c	0.964	2.85 ± 0.23 e	0.63 ± 0.05 c	0.61 ± 0.01 ab	0.998	2926 ± 292 d
100:12	5.29 ± 1.96 d	0.31 ± 0.16 ab	0.73 ± 0.09 c	0.974	2.07 ± 0.37 c	0.61 ± 0.07 c	0.61 ± 0.01 abc	0.998	2856 ± 214 d
100:30	2.19 ± 0.37 bc	0.65 ± 0.11 ef	0.58 ± 0.03 ab	0.989	1.18 ± 0.14 b	0.49 ± 0.01 c	0.61 ± 0.01 abc	0.998	2799 ± 230 d
100:60	0.91 ± 0.32 ab	0.62 ± 0.03 ef	0.54 ± 0.02 ab	0.998	0.58 ± 0.15 a	0.35 ± 0.03 b	0.62 ± 0.01 bc	0.998	1719 ± 146 c
100:120	0.37 ± 0.1 a	0.39 ± 0.06 bc	0.56 ± 0.02 ab	0.999	0.27 ± 0.02 a	0.18 ± 0.01 a	0.66 ± 0.01 d	0.999	1207 ± 176 b

The measurements were performed in triplicate. The mean ± SD values within the same column for each sample, followed by different lowercase letters, are significantly different ($p < 0.05$).

2.4. FT-IR

The infrared spectra of WS and WS−PA gels are displayed in Figure 5a. Only one full IR spectrum of sample with ratio of 100:0 (WS:PA, w/w) is shown, and the others are displayed in Figure S1. The detailed IR spectrums of all samples in the wavenumber range of 1200–800 cm^{-1} are shown in the Figure 5a. All curves exhibit a noticeable absorption peak in the range of 3600–3000 cm^{-1}, corresponding to the stretching vibrations of O-H bonds in starch [33]. The symmetric and antisymmetric stretching vibrations of -CH$_2$ and -CH groups are detected at 3000–2800 cm^{-1} [29]. A minor absorption peak at 1550–1750 cm^{-1} indicates the bending vibration of water molecules within the sample. The fingerprint region of 1200–800 cm^{-1} is sensitive to conformation of the samples, particularly the absorption peaks at 1022 cm^{-1} and 1047 cm^{-1}, which corresponds to the relative contents of random structures in the amorphous region and the ordered structures in the crystalline region, respectively. The ratio of the absorption peak intensity at 1047 cm^{-1} to 1022 cm^{-1} serves as a measure of starch crystallization degree. The results are presented in Figure 5b. It is observed that the ratio of the starch absorption peak at 1047 cm^{-1} to 1022 cm^{-1} increases with the addition of PA, further indicating that during the storage, PA within the system can effectively promote the ordered rearrangement of starch molecular chains.

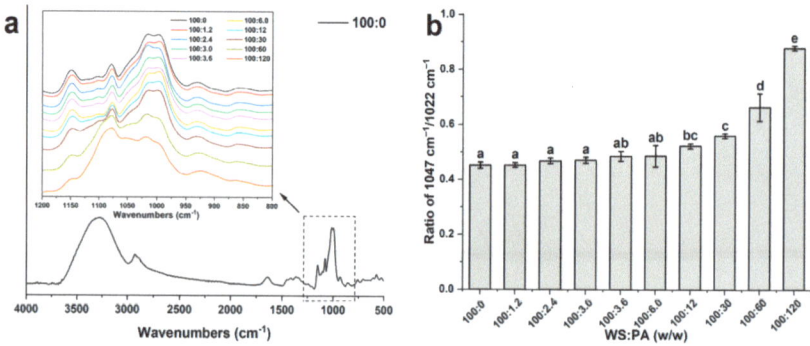

Figure 5. (a) FT-IR spectra of WS and WS−PA gels; (b) the ratio of absorption at 1047 cm^{-1} and 1022 cm^{-1}. The measurements were performed in triplicate. The mean ± SD values within the same column for each sample, followed by different lowercase letters, are significantly different ($p < 0.05$).

2.5. SEM

Figure 6 shows the SEM images of WS and WS−PA gels, magnified 500 times. The microstructure of WS and WS−PA gels resemble a honeycomb structure. However, compared to the control, the gels with addition of PA exhibit larger holes, and the size of the pores increases with higher PA concentrations. The pores in starch gels represent areas where water is distributed during storage. The larger pores and thicker walls in the gels indicate an enhancement of the entanglement force between starch chains generating the separation of water from the homogeneous gel phase [27]. As a result, the addition of PA facilitates the interaction between starch molecules. Furthermore, the structure becomes more uniform and regular and the size of pores grows larger with the increasing ratios of WS:PA from 100:0 to 100:6.0 (w/w). However, the opposite tendency can be observed at the higher ratios from 100:6.0 to 100:120 (w/w). The addition of PA promotes the breakage of starch granules below the ratios of 100:6.0 (WS:PA, w/w), and more starch molecules participate in the network formation of WS−PA gels. However, when the ratios are above 100:6.0 (WS:PA, w/w), the addition of PA inhibits the disintegration of granules. Hence, the structure becomes irregular and the size of pores grows smaller. Additionally, the lyophilized samples with the ratios above 100:30 (WS:PA, w/w) are brittle, which is perhaps related with the high crystallization degree of starch gels.

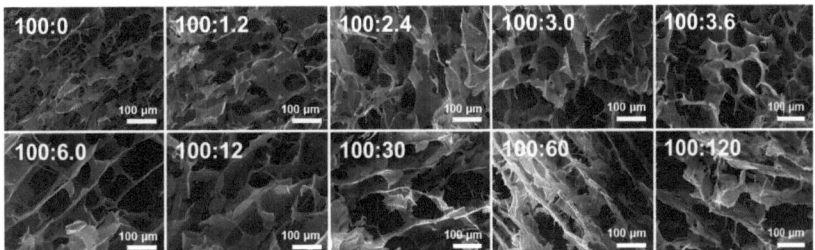

Figure 6. The SEM images of WS and WS−PA gels.

2.6. Thermogravimetry (TG)

Figure 7 shows the thermal degradation analysis of the lyophilized samples using TGA (thermogravimetric analysis). The thermal decomposition curves can be categorized into three main stages, providing information about the thermal stability of samples [34].The first stage (30 °C–150 °C) is related to the release of water and volatile substances [35]. The second stage (200 °C–335 °C) where the main weight loss occurs corresponds to the thermal decomposition of starch [36]. The third stage (400 °C–600 °C) is primarily attributed to the degradation of glycosidic bonds and further decomposition of polymer fragments. The addition of PA promotes the thermal degradation of starch molecules. Compared to the contrast group, the samples mixed with PA contain more moisture. This result can be attributed to the hydrolysis of Al^{3+}, which generates various hydroxyl coordination compounds. These compounds can participate in the network formation of the starch gel. Additionally, a significant amount of hydrogen ions is generated during the hydrolysis process of Al^{3+}, which facilitates the breakage of glycosidic bonds [37]. As a consequence, the temperature range of the second stage becomes narrower and lower when PA is added. Additionally, the addition of PA may also participate in the cross-linking reaction of starch fragments, inhibiting further decomposition of the starch. Hence, the third decomposition temperature to shift to a higher temperature. Moreover, the residual weight of the samples increases with the addition of PA, which can be attributed to the growing mass of PA in the samples.

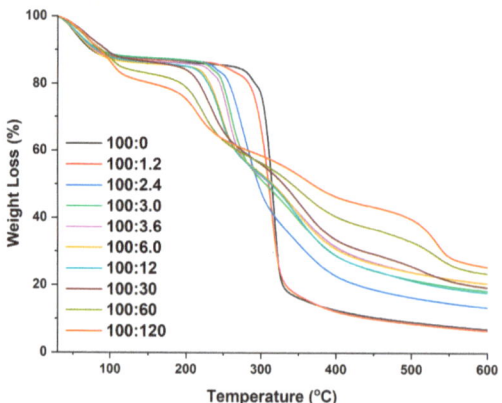

Figure 7. The TGA curves of WS and WS−PA gels.

2.7. Adhesive Properties

2.7.1. Mechanical Properties

Figure 8 shows the mechanical properties of paper samples adhered by either WS or WS−PA gels. The mechanical testing results are directly related to the adhesive strength of the starch gels. There are two main factors that influence the mechanical properties of the testing samples. Firstly, the apparent viscosity and modulus of the starch gels play a significant role. These properties are closely related to the entanglement of starch chains and the ratio of ordered region to the amorphous structure [32]. The starch gels provide strength to the bonding layer between two paper samples. Secondly, the bonding strength between the starch and cellulose fibers on the paper surface is important. This factor is influenced by the permeability of the starch gels and the entanglement between the starch fragments and cellulose fibers. The addition of PA at ratios ranging from 100:0 to 100:120 (WS:PA, w/w) resulted in an increase in folding endurance from 1.96 ± 0.18 to 2.49 ± 0.13, tearing strength from 267.33 ± 15.81 (mN) to 417.66 ± 20.77 (mN), and tensile strength from 0.96 ± 0.08 (kN·m^{-1}) to 1.18 ± 0.10 (kN·m^{-1}). This indicates that the adhesive strength of starch gels was enhanced with the addition of PA. Interestingly, these results were not consistent with the changes observed in the apparent viscosity measured through steady rheology experiments. The reason for this phenomenon can be attributed to the positively charged hydroxyl coordination compounds of Al^{3+}, which act as ion-bridges to connect the starch fragments and cellulose fibers of the paper [24,38]. These ion-bridges provide stronger bonding strength than other influencing factors. This explanation is further supported by the breaking elongation testing results shown in Figure 8d. The minimum elongation was observed at the ratio of 100:120 (WS:PA, w/w), indicating that the higher crosslinked density between cellulose fibers and starch fragments improved the hardness of the paper samples [39]. However, the mechanical properties of the testing samples greatly decreased after the aging experiments at 105 °C for 28 days, and the decline proportion increased with increasing PA addition concentrations. Specifically, the folding endurance, tearing strength, and tensile strength of the control group only decreased by 1.5%, 1.7%, and 1.4%, respectively. In contrast, the corresponding mechanical properties of paper samples adhered by WS−PA gels at a ratio of 100:120 (WS:PA, w/w) decreased by 19.9%, 37%, and 11.5%, respectively. The reason for this phenomenon can be related to the hydrolyzation of Al^{3+} ion, and a large number of H$^+$ ions are generated during this process, which promotes the acid hydrolysis of glycosidic bonds in both starch chains and cellulose fibers. The decrease in mechanical properties of the testing samples after aging experiments highlights the significance of controlling the addition of PA to the starch gel in practical applications, as it can affect the durability and stability of the paper relics.

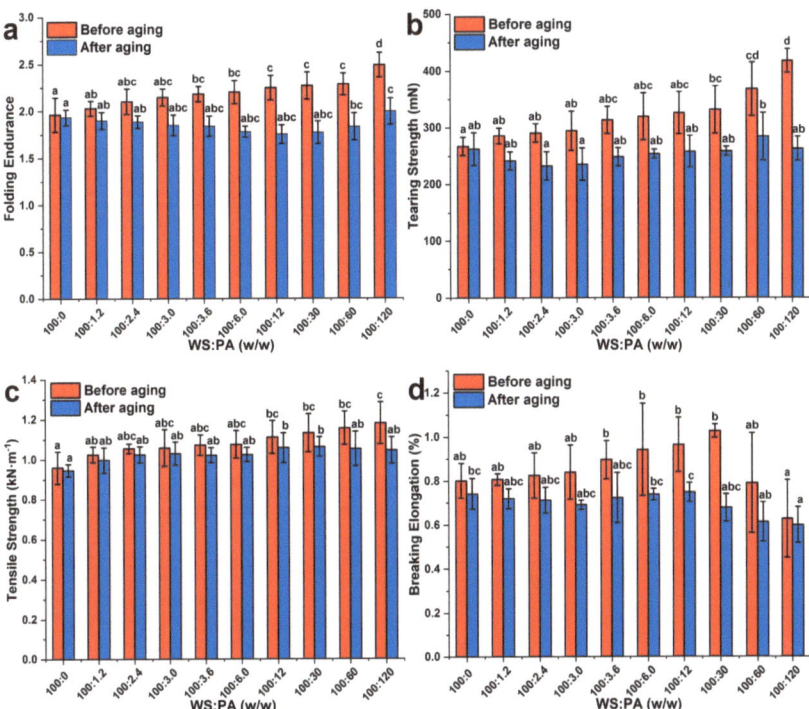

Figure 8. Effect of WS and WS–PA gels on the (**a**) tearing strength, (**b**) folding endurance, (**c**) tensile strength, and (**d**) breaking elongation of paper samples before or after aging experiments. The measurements were performed in triplicate. The mean ± SD values within the same column for each sample, followed by different lowercase letters, are significantly different ($p < 0.05$).

2.7.2. Chromaticity Measurement

Table 3 presents the results of chromaticity measurements. After aging experiments, the color difference of the testing samples is noticeable. Specifically, as the concentrations of PA added into the starch gels increase, the surface of the paper samples becomes darker, redder, and more yellowish in appearance. This change in color can be attributed to the formation of chromophores resulting from the acid degradation of polysaccharides. Acidolysis poses a significant challenge for paper relics, particularly those made from mechanical wood pulp in the 20th century, which were often produced using alum-rosin sizing [40]. These paper relics are known to be brittle, yellowed, and have low pH values. The effect of aging on the color of paper samples is an important consideration in the restoration and preservation of paper relics, as it can impact their aesthetic appearance and historical value.

Table 3. The CIE Color Coordinates of paper samples coated with WS or WS–PA gels after aging at 105 °C for 28 days.

WS:PA (w/w)	L*	a*	b*	ΔE
blank	91.81	−0.24	2.93	
100:0	90.67 ± 0.06 g	−0.2 ± 0.02 a	3.18 ± 0.07 a	1.17 ± 0.04 a
100:1.2	90.53 ± 0.04 g	−0.17 ± 0.04 ab	3.24 ± 0.09 a	1.32 ± 0.02 b
100:2.4	89.73 ± 0.11 f	−0.16 ± 0.07 ab	3.3 ± 0.11 ab	2.12 ± 0.09 c
100:3.0	89.68 ± 0.09 ef	−0.07 ± 0.04 bc	3.4 ± 0.12 b	2.19 ± 0.06 c
100:3.6	89.56 ± 0.08 de	−0.09 ± 0.05 c	3.74 ± 0.04 c	2.40 ± 0.06 d
100:6.0	89.48 ± 0.04 d	0.01 ± 0.03 d	4.05 ± 0.06 d	2.60 ± 0.01 e
100:12	89.19 ± 0.12 c	0.03 ± 0.03 de	4.3 ± 0.1 e	2.97 ± 0.06 f
100:30	89.08 ± 0.05 c	0.07 ± 0.02 def	6.93 ± 0.12 f	4.85 ± 0.07 g
100:60	88.9 ± 0.1 b	0.11 ± 0.07 ef	7.07 ± 0.05 f	5.07 ± 0.01 h
100:120	88.6 ± 0.11 a	0.13 ± 0.06 f	7.43 ± 0.07 g	5.54 ± 0.01 i

The measurements were performed in triplicate. The mean ± SD values within the same column for each sample, followed by different lowercase letters, are significantly different ($p < 0.05$).

3. Materials and Methods

3.1. Materials

Wheat starch (amylose content: 24.3%, water content: 9.6%) was provided by Shanghai Yuanye Bio-Technology Co., Ltd., Shanghai, China. Potassium alum (PA) was provided by Sinopharm. Deionized water was used in this work. Xuan paper was obtained from China Xuan Paper Company Group, Jingxian, China.

3.2. WS–PA Systems Preparation

Different concentrations of PA solutions were prepared. Wheat starch was added into PA solutions to obtain 4% wheat starch solutions by oscillating. Samples with ten ratios of WS:PA (w/w), including 100:0, 100:1.2, 100:2.4, 100:3.0, 100:3.6, 100:6.0, 100:12, 100:30, 100:60, and 100:120, were labeled and used for further experiments. The starch gels were prepared by heating at 95 °C for 20 min under constant stirring. After heating, the resulting starch gels were stored at ambient temperature for 24 h.

3.3. Rheological Measurements

The rheological measurements were carried out by a rheometer (HAAKE MARS III, Thermofisher, Waltham, MA, USA). The experiments were conducted at a temperature of 25 °C using a parallel plate (35 mm diameter, 0.5 mm gap). Before testing, the samples were transferred onto the rheometer plate and equilibrated for 5 min. In order to analyze the dynamic rheological properties, the liner viscoelastic region was determined by a strain sweep measurement, and the strain for the frequency sweep experiment was set at 1% [41]. The frequency range was set from 0.1 to 10 Hz, and the dynamic modulus as functions of frequency was obtained. Steady shear experiments were conducted to measure the response of the samples to varying shear rates. The shear rate was increased from $0.1~\text{s}^{-1}$ to $1000~\text{s}^{-1}$ in 2 min and then decreased from $1000~\text{s}^{-1}$ back to $0.1~\text{s}^{-1}$ at the same speed. The apparent viscosity of WS–PA gels was recorded as a function of shear rate. To analyze the results of the steady shear tests, the data were fitted using the Herschel–Bulkley model, which is denoted by Equation (1) [42].

$$\tau = \tau_0 + Kx^n \tag{1}$$

where τ_0 represents the yield stress (Pa), τ represents shear stress (Pa), K is consistency index (Pa·s^n), γ represents shear rate (s^{-1}), and n is the flow behavior coefficient.

3.4. Leached Amylose and Swelling Power

The leached amylose and swelling power of WS and WS−PA gels were obtained according to the report of Chen et al. with a slight modification [43]. The slurries were prepared as per the descriptions of Section 3.2 and then the mixtures heated in an oscillating water bath at temperatures of 75 °C, 85 °C, and 95 °C for a duration of 30 min. After heating, the samples were immediately cooled to ambient temperature, and then the cooled samples were centrifuged at 4800 rpm for 30 min. The sediment was weighed and then dried to constant weight at 90 °C. The swelling power (%) was denoted as the ratio between the weight of the precipitate in its wet state and its dry weight. The amount of leached amylose in the supernatant were determined using the iodine colorimetric method as reported by Chrastil with a slight modification [44]. In detail, 0.05 mL of the supernatant obtained after centrifugation was added to 3 mL of 0.33 M NaOH in a water bath set at 95 °C for 30 min. After cooling to ambient temperature, the mixtures were centrifuged at 4800 rpm for 10 min. Then, 0.1 mL of the resulting solutions was mixed with 2.5 mL of trichloroacetic acid (0.5%, v/v) to adjust the pH of the solution to 5.0–6.0, and subsequently 0.01 N I_2-KI (0.05 mL) was added. The mixture was reacted at room temperature for 30 min, and the absorbance of solution at 620 nm was measured using a spectrophotometer (TGem Plus, TIANGEN, Beijing, China). Amylose was used as the standard to calculate the dissolution amount of amylose.

3.5. DSC

The thermal analysis of wheat starch with different concentrations of PA was conducted using a DSC (DSC 250, TA Instruments, New Castle, DE, USA) under a N_2 atmosphere [4]. To perform the analysis, 3 mg of wheat starch was mixed with 6 μL of various concentrations of PA solutions. This mixture was then hermetically sealed in an aluminum pan. To ensure full hydration, all samples were equilibrated at room temperature for 12 h. The temperature range for the analysis was set from 35 °C to 95 °C, and the heating rate was 10 °C/min. The DSC equipment software (TRIOS software version 5.4.0.300) was used to calculate several parameters of interest, including the onset temperature (To), peak temperature (Tp), conclude temperature (Tc), and the area of the main endothermic peak (J/g).

3.6. TGA

The WS−PA gels were prepared according to the methods in Section 3.2 and stored at −80 °C for 24 h. Subsequently, the samples were subjected to freeze–drying using a vacuum freeze dryer. To evaluate the thermal stability of the WS−PA samples, a thermal gravimetric analyzer (TGA, Mettler Toledo, Zurich, Switzerland) was employed. The experiments were carried out under a N_2 atmosphere, and the heating rate was 20 °C/min. The temperature range was set from 30 °C to 600 °C. The weight loss of the samples was recorded as a function of temperature, indicating the decomposition or volatilization of different components contained in the samples [34,45].

3.7. ATR-FTIR

The lyophilized samples were obtained following the procedure described in Section 3.6. The measurements were conducted using an FT-IR instrument (Nicolet 6700, Thermofisher, Waltham, MA, USA) equipped with a diamond ATR accessory over the range from 4000 cm^{-1} to 500 cm^{-1}. To collect the spectra, each sample was tested at a resolution of 4 cm^{-1}, with an accumulation of 64 scans. One specific parameter of interest for evaluating the molecular order of starch gels is the ratio of absorbances at 1047 cm^{-1} and 1022 cm^{-1}. This ratio provides an indication of the short-range molecular order present in the starch gels. By comparing the intensity of these two absorbance peaks, it is possible to evaluate the structural characteristics and organization of the starch gel [46].

3.8. SEM

The lyophilized samples were obtained following the procedure described in Section 3.6. The microstructure of the lyophilized samples was observed using SEM (VEGA 3 XMU, TESCAN, Brno, Czech) [47]. The accelerating beam voltage was set at 20 kV. To prepare the samples for SEM analysis, the cross-section of the lyophilized samples was coated with a layer of gold. After the gold coating, the samples were placed on the loading platform of the SEM using conductive adhesive tape. By utilizing SEM, high-resolution images of the sample's surface can be obtained.

3.9. Paper Mechanical Properties Experiments

The starch gels were obtained as per the methods described in Section 3.2. The gels were coated evenly on one piece of paper (0.25 m × 0.25 m) by a coir scrub brush, and then another piece of paper was put on it and adhered together. After dried at 30 °C for 24 h, the paper samples were cut into different sizes for further mechanical properties experiments.

3.9.1. Tensile Strength Experiment

The tensile strength was measured according to GB/T 12914-2008 at a constant elongation rate of 20 mm/min by a tensile strength tester (ZB-WLQ, Hangzhou Zhibang Automation Technology Co., Ltd., Hangzhou, China), and the size of paper samples was 15 cm × 0.75 cm.

3.9.2. Folding Endurance Experiment

The folding endurance was measured according to GB/T 475-2008 by an MIT folding endurance tester (ZB-NZ135A, Hangzhou Zhibang Automation Technology Co., Ltd., Hangzhou, China). The size of paper samples was 15 cm × 1.5 cm, and the force was set at 4.98 N.

3.9.3. Tearing Strength Experiment

The tearing strength was performed according to GB/T 455-2002 by a tearing strength tester (ZB-SL, Hangzhou Zhibang Automation Technology Co., Ltd., Hangzhou, China). The size of paper samples was 63 cm × 50 cm.

3.10. Color Evaluation

The color change of paper samples before and after the aging process was determined by the color coordinate values (Equation (2)) using an NR10QC colorimeter [48].

$$\Delta E = [(\Delta L^*)^2 + (\Delta a^*)^2 + (\Delta b^*)^2]^{1/2} \tag{2}$$

where "L^*" represents brightness, "a^*" represents red and green, and "b^*" represents yellow and blue.

3.11. Statistical Analysis

All the measurements were conducted in triplicate to ensure accuracy and reliability of the results. The results were subjected to statistical analysis using SPSS statistical software (Version 20.0, SPSS Inc., Chicago, IL, USA). To assess the significance, analysis of variance (ANOVA) was performed on the data. In this case, Duncan's test was used to further analyze the data, with a significance level set at $p < 0.05$. Additionally, the graphical representation of the data was created using Origin Pro software, specifically Version 8.0 by Stat-Ease Inc. (Minneapolis, MN, USA).

4. Conclusions

The addition of potassium alum to wheat starch gels was found to be an effective method for modifying their physicochemical properties. The experimental results show that addition of PA increases the gelatinization temperature and enthalpy value of the starch gels. However, it was observed that the leached amylose and swelling power of the starch gels reached a maximum at a certain ratio of WS:PA, specifically at the ratio of 100:6.0 (w/w). The rheological properties and SEM observations support this phenomenon. In practical applications, the addition of PA enhanced the bonding strength of the starch gels. The adhesive ability of the starch gels was improved with addition of PA. However, the addition of PA at high concentrations could have a negative effect on the surface color of paper samples. Therefore, it is advisable to carefully control the concentration of PA when adding it to starch gels. Keeping the addition of PA at low concentrations, specifically below the ratio of 100:6.0 (WS:PA, w/w), could help to mitigate the negative effects of PA on the degradation and color change of paper relics while still achieving the desired enhancements in rheological and adhesive properties.

Supplementary Materials: The following supporting information can be downloaded at: https://www.mdpi.com/article/10.3390/molecules28186670/s1, Figure S1: FT-IR spectra of WS and WS–PA gels.

Author Contributions: Conceptualization, H.Z. (Haibo Zhao) and Q.X.; methodology, H.Z. (Haibo Zhao); software, H.Z. (Haibo Zhao); validation, H.Z. (Haibo Zhao) and H.Z. (Hongbin Zhang); formal analysis, H.Z. (Haibo Zhao) and Q.X.; investigation, H.Z. (Haibo Zhao); resources, H.Z. (Hongdong Zhang) and Y.Y.; data curation, H.Z. (Haibo Zhao) and H.Z. (Hongbin Zhang); writing—original draft preparation, H.Z. (Haibo Zhao); writing—review and editing, H.Z. (Haibo Zhao), H.Z. (Hongbin Zhang), H.Z. (Hongdong Zhang) and Y.Y.; supervision, H.Z. (Hongdong Zhang) and Y.Y.; project administration, H.Z. (Hongdong Zhang) and Y.Y.; funding acquisition, H.Z. (Hongdong Zhang) and Y.Y. All authors have read and agreed to the published version of the manuscript.

Funding: This research was funded by the Institute for Preservation and Conservation of Chinese Ancient Books (No. SCH6004101).

Institutional Review Board Statement: Not applicable.

Informed Consent Statement: Not applicable.

Data Availability Statement: Data may be shared under request.

Conflicts of Interest: The authors declare no conflict of interest. The funders had no role in the design of the study; in the collection, analyses, or interpretation of data; in the writing of the manuscript; or in the decision to publish the results.

Sample Availability: Samples of the compounds are available from the authors.

References

1. Biricik, Y.; Sonmez, S.; Ozden, O. Effects of surface sizing with starch on physical strength properties of paper. *Asian J. Chem.* **2011**, *23*, 3151–3154.
2. Yu, J.; Wang, S.; Wang, J.; Li, C.; Xin, Q.; Huang, W.; Zhang, Y.; He, Z.; Wang, S. Effect of laboratory milling on properties of starches isolated from different flour millstreams of hard and soft wheat. *Food Chem.* **2015**, *172*, 504–514. [CrossRef] [PubMed]
3. Li, H.; Qi, Y.; Zhao, Y.; Chi, J.; Cheng, S. Starch and its derivatives for paper coatings: A review. *Prog. Org. Coat.* **2019**, *135*, 213–227. [CrossRef]
4. Chen, L.; Ren, F.; Zhang, Z.; Tong, Q.; Rashed, M.M. Effect of pullulan on the short-term and long-term retrogradation of rice starch. *Carbohydr. Polym.* **2015**, *115*, 415–421. [CrossRef]
5. Wang, Y.; Zhan, J.; Lu, H.; Chang, R.; Qiu, L.; Tian, Y. Amylopectin crystal seeds: Characterization and their effect on amylopectin retrogradation. *Food Hydrocoll.* **2021**, *111*, 106409. [CrossRef]
6. Ma, S.; Zhu, P.; Wang, M. Effects of konjac glucomannan on pasting and rheological properties of corn starch. *Food Hydrocoll.* **2019**, *89*, 234–240. [CrossRef]
7. Singh, A.; Geveke, D.J.; Yadav, M.P. Improvement of rheological, thermal and functional properties of tapioca starch by using gum arabic. *LWT* **2017**, *80*, 155–162. [CrossRef]
8. Lian, C. Study on the Viscose for the Archives Decoration and Mount. *J. Fujian Norm. Univ.* **1999**, *3*, 63–67.
9. Cao, F.; Liu, S.; Zheng, L.; Yu, J. Application of starch adhesive in mounting process of traditional Chinese painting and calligraphy. *Adhesion* **2008**, *5*, 47–50.
10. Sasaki, A.; Kishigami, Y.; Fuchigami, M. Firming of cooked sweet potatoes as affected by alum treatment. *J. Food Sci.* **1999**, *64*, 111–115. [CrossRef]

11. Baranova, L.I.; Syrkin, Y.K.; Kuchumova, L.M. Dielectric polarization of water in aluminum-potassium alums. *J. Struct. Chem.* **1966**, *7*, 467–469. [CrossRef]
12. Chen, Y.; Wang, C.; Chang, T.; Shi, L.; Yang, H.; Cui, M. Effect of salts on textural, color, and rheological properties of potato starch gels. *Starch Stärke* **2014**, *66*, 149–156. [CrossRef]
13. Li, W.; Bai, Y.; Zhang, Q.; Hu, X.; Shen, Q. Effects of potassium alum addition on physicochemical, pasting, thermal and gel texture properties of potato starch. *Int. J. Food Sci. Technol.* **2011**, *46*, 1621–1627. [CrossRef]
14. Li, Z.; Zhang, Y.; Ai, Z.; Fan, H.; Wang, N.; Suo, B. Effect of potassium alum addition on the quality of potato starch noodles. *J. Food Sci. Technol.* **2019**, *56*, 2932–2939. [CrossRef] [PubMed]
15. Wang, W.; Zhou, H.; Yang, H.; Zhao, S.; Liu, Y.; Liu, R. Effects of salts on the gelatinization and retrogradation properties of maize starch and waxy maize starch. *Food Chem.* **2017**, *214*, 319–327. [CrossRef] [PubMed]
16. Kang, B.; Tang, H.; Zhao, Z.; Song, S. Hofmeister Series: Insights of Ion Specificity from Amphiphilic Assembly and Interface Property. *ACS Omega* **2020**, *5*, 6229–6239. [CrossRef]
17. Zhu, W.X.; Gayin, J.; Chatel, F.; Dewettinck, K.; Van der Meeren, P. Influence of electrolytes on the heat-induced swelling of aqueous dispersions of native wheat starch granules. *Food Hydrocoll.* **2009**, *23*, 2204–2211. [CrossRef]
18. Jane, J. Mechanism of Starch Gelatinization in Neutral Salt Solutions. *Starch Stärke* **1993**, *45*, 161–166. [CrossRef]
19. Beck, M.; Jekle, M.; Becker, T. Starch re-crystallization kinetics as a function of various cations. *Starch Stärke* **2011**, *63*, 792–800. [CrossRef]
20. Zervos, S.; Alexopoulou, I. Paper conservation methods: A literature review. *Cellulose* **2015**, *22*, 2859–2897. [CrossRef]
21. Liu, J.; Xing, H.; Zhou, Y.; Chao, X.; Li, Y.; Hu, D. An Essential Role of Polymeric Adhesives in the Reinforcement of Acidified Paper Relics. *Polymers* **2022**, *14*, 207. [CrossRef] [PubMed]
22. Huang, C.-C. Physicochemical, pasting and thermal properties of tuber starches as modified by guar gum and locust bean gum. *Int. J. Food Sci. Technol.* **2009**, *44*, 50–57. [CrossRef]
23. Nguyen, Q.D.; Jensen, C.T.B.; Kristensen, P.G. Experimental and modelling studies of the flow properties of maize and waxy maize starch pastes. *Chem. Eng. J.* **1998**, *70*, 165–171. [CrossRef]
24. Rosenholm, J.B. Critical evaluation of dipolar, acid-base and charge interactions II. Charge exchange within electrolytes and electron exchange with semiconductors. *Adv. Colloid. Interface Sci.* **2017**, *247*, 305–353. [CrossRef]
25. Xie, J.; Ren, Y.; Xiao, Y.; Luo, Y.; Shen, M. Interactions between tapioca starch and Mesona chinensis polysaccharide: Effects of urea and NaCl. *Food Hydrocoll.* **2021**, *111*, 106268. [CrossRef]
26. Luo, Y.; Shen, M.; Han, X.; Wen, H.; Xie, J. Gelation characteristics of Mesona chinensis polysaccharide-maize starches gels: Influences of KCl and NaCl. *J. Cereal Sci.* **2020**, *96*, 103108. [CrossRef]
27. Luo, Y.; Shen, M.; Li, E.; Xiao, Y.; Wen, H.; Ren, Y.; Xie, J. Effect of Mesona chinensis polysaccharide on pasting, rheological and structural properties of corn starches varying in amylose contents. *Carbohydr. Polym.* **2020**, *230*, 115713. [CrossRef]
28. Zhou, H.; Wang, C.; Shi, L.; Chang, T.; Yang, H.; Cui, M. Effects of salts on physicochemical, microstructural and thermal properties of potato starch. *Food Chem.* **2014**, *156*, 137–143. [CrossRef]
29. Ren, Y.; Rong, L.; Shen, M.; Liu, W.; Xiao, W.; Luo, Y.; Xie, J. Interaction between rice starch and Mesona chinensis Benth polysaccharide gels: Pasting and gelling properties. *Carbohydr. Polym.* **2020**, *240*, 116316. [CrossRef]
30. Kong, X.-R.; Zhu, Z.-Y.; Zhang, X.-J.; Zhu, Y.-M. Effects of Cordyceps polysaccharides on pasting properties and in vitro starch digestibility of wheat starch. *Food Hydrocoll.* **2020**, *102*, 105604. [CrossRef]
31. Wagner, C.E.; Barbati, A.C.; Engmann, J.; Burbidge, A.S.; McKinley, G.H. Apparent shear thickening at low shear rates in polymer solutions can be an artifact of non-equilibration. *Appl. Rheol.* **2016**, *26*, 54091.
32. Liu, S.; Lin, L.; Shen, M.; Wang, W.; Xiao, Y.; Xie, J. Effect of Mesona chinensis polysaccharide on the pasting, thermal and rheological properties of wheat starch. *Int. J. Biol. Macromol.* **2018**, *118*, 945–951. [CrossRef] [PubMed]
33. Zhao, D.; Deng, Y.; Han, D.; Tan, L.; Ding, Y.; Zhou, Z.; Xu, H.; Guo, Y. Exploring structural variations of hydrogen-bonding patterns in cellulose during mechanical pulp refining of tobacco stems. *Carbohydr. Polym.* **2019**, *204*, 247–254. [CrossRef]
34. Ren, Y.; Wu, Z.; Shen, M.; Rong, L.; Liu, W.; Xiao, W.; Xie, J. Improve properties of sweet potato starch film using dual effects: Combination Mesona chinensis Benth polysaccharide and sodium carbonate. *LWT* **2021**, *140*, 110679. [CrossRef]
35. Rong, L.; Shen, M.; Wen, H.; Xiao, W.; Li, J.; Xie, J. Effects of xanthan, guar and Mesona chinensis Benth gums on the pasting, rheological, texture properties and microstructure of pea starch gels. *Food Hydrocoll.* **2022**, *125*, 107391. [CrossRef]
36. Valencia-Sullca, C.; Atarés, L.; Vargas, M.; Chiralt, A. Physical and Antimicrobial Properties of Compression-Molded Cassava Starch-Chitosan Films for Meat Preservation. *Food Bioprocess Technol.* **2018**, *11*, 1339–1349. [CrossRef]
37. Zhang, H.; Zhang, C.; Ye, Z.; Wang, S.; Tang, Y. Alkali-exchanged Y zeolites as superior deacidifying protective materials for paper relics: Effects of accessibility and strength of basic sites. *Microporous Mesoporous Mater.* **2020**, *293*, 109786. [CrossRef]
38. He, Q.; Wang, L.; Zhang, Y. Study of Mechanism of Aluminum Sizing Precipitant on Xuan Paper Based on Spectral Analysis. *Spectrosc. Spectr. Anal.* **2018**, *38*, 418–423.
39. Chen, G.; Zhu, Z.J.; Salminen, P.; Toivakka, M. Structure and Mechanical Properties of Starch/Styrene-Butadiene Latex Composites. *Adv. Mater. Res.* **2014**, *936*, 74–81. [CrossRef]
40. Wang, F.; Wu, Z.; Tanaka, H. Preparation and sizing mechanisms of neutral rosin size II: Functions of rosin derivatives on sizing efficiency. *J. Wood Sci.* **1999**, *45*, 475–480. [CrossRef]

41. Gałkowska, D.; Pycia, K.; Juszczak, L.; Pająk, P. Influence of cassia gum on rheological and textural properties of native potato and corn starch. *Starch Stärke* **2014**, *66*, 1060–1070. [CrossRef]
42. Dangi, N.; Yadav, B.S.; Yadav, R.B. Pasting, rheological, thermal and gel textural properties of pearl millet starch as modified by guar gum and its acid hydrolysate. *Int. J. Biol. Macromol.* **2019**, *139*, 387–396. [CrossRef] [PubMed]
43. Chen, L.; Tong, Q.; Ren, F.; Zhu, G. Pasting and rheological properties of rice starch as affected by pullulan. *Int. J. Biol. Macromol.* **2014**, *66*, 325–331. [CrossRef]
44. Chrastil, J. Improved colorimetric determination of amylose in starches or flours. *Carbohydr. Res.* **1987**, *159*, 154–158. [CrossRef]
45. Ali, A.; Xie, F.; Yu, L.; Liu, H.; Meng, L.; Khalid, S.; Chen, L. Preparation and characterization of starch-based composite films reinfoced by polysaccharide-based crystals. *Compos. Part B Eng.* **2018**, *133*, 122–128. [CrossRef]
46. Guo, P.; Yu, J.; Copeland, L.; Wang, S.; Wang, S. Mechanisms of starch gelatinization during heating of wheat flour and its effect on in vitro starch digestibility. *Food Hydrocoll.* **2018**, *82*, 370–378. [CrossRef]
47. Krstonosic, V.; Dokic, L.; Nikolic, I.; Milanovic, M. Influence of xanthan gum on oil-in-water emulsion characteristics stabilized by OSA starch. *Food Hydrocoll.* **2015**, *45*, 9–17. [CrossRef]
48. Ma, X.; Zhu, Z.; Zhang, H.; Tian, S.; Li, X.; Fan, H.; Fu, S. Superhydrophobic and deacidified cellulose/CaCO(3)-derived granular coating toward historic paper preservation. *Int. J. Biol. Macromol.* **2022**, *207*, 232–241. [CrossRef]

Disclaimer/Publisher's Note: The statements, opinions and data contained in all publications are solely those of the individual author(s) and contributor(s) and not of MDPI and/or the editor(s). MDPI and/or the editor(s) disclaim responsibility for any injury to people or property resulting from any ideas, methods, instructions or products referred to in the content.

Article

An Investigation into the Performance and Mechanisms of Soymilk-Sized Handmade Xuan Paper at Different Concentrations of Soymilk

Chunfang Wu [1], Yangyang Liu [2], Yanxiao Hu [1], Ming Ding [3,4], Xiang Cui [5], Yixin Liu [5], Peng Liu [1], Hongbin Zhang [1,*], Yuliang Yang [1,5,*] and Hongdong Zhang [5,*]

1. Institute for Preservation and Conservation of Chinese Ancient Books, Fudan University, Shanghai 200433, China; 19110820002@fudan.edu.cn (C.W.)
2. School of Creative Art and Fashion Design, Huzhou Vocational and Technical College, Huzhou 313000, China
3. Behavioral and Cognitive Neuroscience Center, Institute of Science and Technology for Brain-Inspired Intelligence, Fudan University, Shanghai 200433, China
4. Department of Rehabilitation Medicine, Huashan Hospital, Fudan University, Shanghai 200040, China
5. State Key Laboratory of Molecular Engineering of Polymers, Department of Macromolecular Science, Fudan University, Shanghai 200433, China
* Correspondence: zhanghongbin@fudan.edu.cn (H.Z.); yuliangyang@fudan.edu.cn (Y.Y.); zhanghd@fudan.edu.cn (H.Z.)

Citation: Wu, C.; Liu, Y.; Hu, Y.; Ding, M.; Cui, X.; Liu, Y.; Liu, P.; Zhang, H.; Yang, Y.; Zhang, H. An Investigation into the Performance and Mechanisms of Soymilk-Sized Handmade Xuan Paper at Different Concentrations of Soymilk. *Molecules* 2023, 28, 6791. https://doi.org/10.3390/molecules28196791

Academic Editor: Rafał M. Łukasik

Received: 15 August 2023
Revised: 7 September 2023
Accepted: 19 September 2023
Published: 25 September 2023

Copyright: © 2023 by the authors. Licensee MDPI, Basel, Switzerland. This article is an open access article distributed under the terms and conditions of the Creative Commons Attribution (CC BY) license (https://creativecommons.org/licenses/by/4.0/).

Abstract: Invaluable paper relics that embody a rich traditional culture have suffered damage, requiring urgent restoration. In this context, the utilization of soymilk as a sizing agent holds great significance and reverence. This study investigates the use of soymilk as a sizing agent for Xuan paper and evaluates its effects on various properties and the long-term behavior of the paper. The findings reveal that the application of soymilk as a sizing agent for Xuan paper imparts distinct properties, including hydrophobicity, improved mechanical properties, and unique chromaticity. These characteristics—arising from the papillae on the surface of the Xuan paper, the protein folding of the soy protein, and hydrogen-bonding interactions between the soy protein and paper fibers—play a crucial role in shaping the paper's unique attributes. From a physicochemical perspective, the aging process leads to multiple changes in paper properties. These changes include acidification, which refers to a decrease in pH, as well as a decline in mechanical strength, an increase in chromaticity, and a decrease in the degree of polymerization (DP) of the paper. The Ekenstam equation is employed to predict the lifespan of the paper, showing longer lifespans for Sheng Xuan paper and a negative correlation between soymilk concentration and lifespan in soymilk-sized paper. Our work provides valuable insights for the preservation and maintenance of paper, highlighting the potential benefits and challenges of using soymilk for surface sizing.

Keywords: surface sizing; soymilk; Xuan paper; lifespan; potential mechanism

1. Introduction

Traditional Chinese calligraphy and painting have always been an important carrier for Chinese literati to express personal emotions or concerns about political life, reflect on social changes, and record historical events, and it is an integral part of the outstanding traditional Chinese culture. From 2012 to 2016, the first national survey of movable cultural relics was carried out under the unified deployment of The State Council of the People's Republic of China. Census data have shown that there are more than 1.5 million paintings and calligraphy cultural relics in China [1]. However, these cultural relics have suffered varying degrees of damage, either due to environmental factors such as armed conflict, conflagration, seismic activity, floods, insolation, and insect infestation, or as a result of human actions like frequent flipping and tearing over time. In order to prevent further damage, most of these cultural relics are facing the urgent need of restoration. Furthermore, the restoration of traditional Chinese paintings and calligraphy has consistently upheld the principle of "Repairing the old as the old" [2]. This principle entails meticulously repairing the damaged sections of cultural artifacts while aiming to restore the original aesthetics of the artwork, ensuring that no visible traces of restoration remain discernible to observers.

During the restoration process of traditional paintings and calligraphy, starch is commonly used as a natural sizing agent. However, natural starch itself has certain limitations, including the high viscosity of starch paste, a poor affinity with fibers, and a tendency to detach easily [3]. These limitations restrict the practical application of natural starch to some extent. In response to these challenges, our ancestors discovered and refined a valuable technique known as the mixed use of alum and gelatin through continuous practice and exploration. When employed in the restoration of calligraphy and paintings, an alum–gelatin solution can effectively secure the pigment on the artwork, preventing it from flaking off. In addition, it enhances the water resistance of the paper. However, it is crucial to find the appropriate ratio of alum to gelatin. If the proportion of alum is too high, it can render the paper cultural relics brittle. The use of an alum–gelatin solution in the restoration of paper cultural relics has been a topic of debate, as there are concerns about its potential harm [4]. Therefore, the demand for safe and harmless sizing agents in cultural relic restoration remains unmet. Soybean, one of the most important global crops, is widely cultivated and consumed worldwide. It contains a multitude of proteins, lipids, and secondary metabolites, granting it a crucial position across various sectors [5]. Hereinto, soybean protein exhibits amphiphilic properties, which have garnered considerable attention in fields such as medicine, beauty, packaging, and food. In particular, soybean protein has also captured significant interest within the domain of cultural relics conservation due to its amphiphilic nature. In the realm of traditional Chinese painting and calligraphy restoration, soybean-derived soymilk has long been utilized as a sizing agent. Esteemed expert Ms. Qiu Jinxian, renowned for her exceptional proficiency in restoring valuable paintings and calligraphy, has contributed significantly to this field during her tenure at both the Shanghai Museum and the British Museum. In her report, Ms. Qiu highlighted that soymilk exhibited remarkable suitability in the restoration of calligraphy and paintings, possessing favorable color properties and hydrophilic and hydrophobic performance. Typically, the recommended concentration is one cup of soymilk with three cups of water [6]. The hydrophilic and hydrophobic properties of soymilk-sized paper can be traced back to the Ming Dynasty. The eighth volume of *Physical Knowledge* written by Fang Yizhi, a Ming Dynasty scholar, documents the process of transforming paper into waterproof window paper with "tofu pulp". In addition, the painter Wu Shanming suggested that the ideal ratio of soymilk to water should be maintained at three to seven [7], which yields the optimal hydrophilic and hydrophobic performance when sizing paper with soymilk.

According to Xu's research, the application of soybean protein as a coating on paper surfaces reduces surface voids and facilitates the formation of hydrogen bonds between protein and fiber hydroxyl groups. This interaction enhances the water resistance of the paper, making soybean protein an effective sizing agent for improving its hydrophobic property [8]. In addition, when soymilk is used as a sizing agent at different temperatures, it is observed that the hydrophilic and hydrophobic performance of the paper can be varied. Specifically, the soybean protein molecules gradually unfold from their tightly coiled structure, which is surrounded by a hydration film, and remain relatively stable. This unfolding exposes the hydrophobic groups within the polypeptide chain's coiled structure while reducing the presence of hydrophilic groups on the outer surface [9]. As the temperature of the soymilk increases from 25 °C to 80 °C, the relative content of the β-sheet structure (one of the secondary structures of protein molecules) increases from 41.14% to 48.87%. Meanwhile, the corresponding α-helix, β-turn, and random coiled structures of soybean protein transform into β-sheet structures, with the proportion of β-sheet structures positively correlated with hydrophobicity. However, the relative content of the β-sheet structure in soybean protein decreases to 45.43% at 90 °C. This reduction in β-sheet structure content indicates that the structure of soybean protein can undergo thermal denaturation at higher temperatures [10]. Consequently, precipitation may occur, compromising the desired properties of the soybean protein sizing agent.

The use of soymilk as a sizing agent for paper either lacks specific information, or the information provides vague details regarding the appropriate ratio of soymilk to water. This study focuses on the influence of the varying concentrations of soymilk on the paper when it is used as a sizing agent. A total of six distinct types of paper were carefully chosen to conduct a comprehensive analysis of the surface properties and the impact of soymilk treatment at both macro and micro levels. In addition, this study also evaluated the lifespan of paper that treated with soymilk. The findings of this research will aid in the selection of optimal soymilk concentrations for paper sizing, thereby establishing a strong foundation for understanding the interaction between paper fibers and soybean protein.

2. Results and Discussion

2.1. Characterization of Soymilk

In Figure 1A, the pH values of soymilk at different concentrations are close to neutral, indicating that they do not pose a threat to paper when used for sizing. The viscosity of soymilk is found to be closely related to its concentration. Higher concentrations of soymilk result in greater viscosity. Herein, it was observed that the 1:5 concentration of soymilk had the lowest viscosity, which was similar to the viscosity of water at ambient temperature (1.01 mPa·s). A lower viscosity indicates strong flowability and easy spreading of the soymilk. This characteristic is beneficial for the soymilk to penetrate into the paper [11]. Therefore, it can be concluded that the experimental concentration of soymilk allows for easy permeation of the paper due to its low viscosity and favorable flowability.

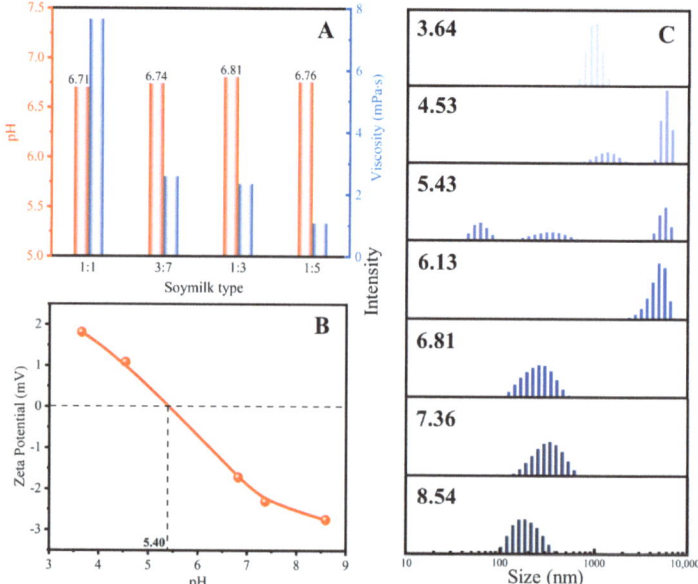

Figure 1. (**A**) pH and viscosity of different concentrations of soymilk, (**B**) zeta potential, and (**C**) size distribution of soymilk at various pH values.

The main ingredient in soymilk is soy proteins, which are composed of albumins and globulins. When extracted by water, albumins represent approximately 10% of soy proteins. Soy protein molecules are complex macromolecules consisting of 20 different amino acids. These amino acids are connected through peptide bonds, forming polypeptide chains, which represent the primary structure of protein. The amino acids' skeleton comprises hydrophilic amino groups, the carboxylic groups, and hydrophobic side chains. Each amino acid, with its specific side chain, plays a unique role in the protein structure [11]. The presence of polar groups and apolar groups in proteins influences the structure in an aqueous solution. Therefore, the surface charge and size distribution of the soy protein were measured under different pH conditions. Figure 1B shows that the zeta potential of soy protein gradually decreases with increasing pH, reaching its isoelectric point at pH 5.40. Many researchers have demonstrated that polysaccharide–protein composite films exhibit improved functional properties. The maximum complexation interaction between protein and polysaccharide, mediated by electrostatic interactions, typically occurs near the isoelectric point. Therefore, the surface potential of soymilk plays a role in its interaction with paper cellulose and can influence paper properties to some extent.

In addition, Figure 1C displays the size distribution of soymilk proteins, revealing that protein particles range in size from 45 nm to 8 μm. Previous research has shown that protein samples can exhibit multiple distributions under certain conditions. Some particles are distributed in the nanoscale range, representing natural protein molecules, while others exist as aggregates of protein molecules and are larger in size [12]. Specifically, as the pH decreases and the proton concentration increases, the amino and carboxylic groups in the protein interact with these protons, resulting in a positively charged surface and a low net charge. At pH 3.64, the net surface charge of protein is +1.82 mV. The repulsion between soy protein and protons leads to a primary protein size distribution ranging from 600 to 1100 nm. With the increase in pH, the proton concentration decreases, weakening the interaction. This change results in a wide range of protein sizes. At pH 4.53, protein sizes exhibit a distribution spanning between 600 to 2000 and 4000 to 8000 nm. As the pH approaches the isoelectric point, polydisperse particles become more noticeable due to reduced electrostatic repulsion and decreased solution stability. The size distribution spans two orders of magnitude at pH 5.43. This aggregation tendency is attributed to the presence of protein molecules with less homogeneous charges on their surfaces [13]. As the pH increases, some amino acids in the protein release protons, leading to a negatively charged environment. The reduction in proton concentration with increasing pH causes an increase in the net negative charge on the protein surface. Electrostatic repulsions between proteins can cause aggregation and a decrease in particle size under high pH conditions [14]. The protein size distribution decreased from 3000~8000 nm at pH 6.13 to 100~600 nm at pH 8.54. It has been reported that the free energy of protein has a strong negative correlation with protein size and microstructure, which, in turn, affects the functionalities [15].

2.2. Analysis of Sized Paper

The quality of paper can be evaluated by basic weight, which considers the basic weight parameter and the influences of physical and printing properties on the paper. In Table 1, the quality of paper coated with different sizing agents is presented. The basic weights of the soymilk-coated paper are higher compared with Sheng Xuan paper, indicating an improvement in quality. The basic weight of soymilk-sized paper decreases as the concentration of soymilk decreases. This means that lower concentrations of soymilk result in a decrease in the weight of the paper. In addition, the basic weight of the soymilk-sized paper is lower compared with paper coated with an alum–gelatin solution. The results indicate a distinct weight advantage of the soymilk-coated paper over Shu Xuan paper, suggesting that soymilk sizing can provide a lighter-weight paper option while maintaining or even improving quality.

Table 1. Basic weights of different papers.

Paper Type	Basic Weight (g/m^2)
Sheng	21.31
Shu	30.44
1:1	27.94
1:3	25.98
3:7	24.70
1:5	22.55

Investigating the dynamic behavior of liquids on surfaces is crucial for understanding the surface properties of Xuan paper and its practical applications in writing and printing. Typically, a water-wetting threshold of 90° is considered to divide surfaces into hydrophilic and hydrophobic categories. Figure 2A shows that the static contact angle of the soymilk-sized paper is significantly improved compared with Sheng Xuan paper and is comparable to that of Shu Xuan. This indicates that the soymilk-sized paper possesses a more hydrophobic surface compared with the uncoated Sheng Xuan paper. Notably, the contact angles of the 3:7 and 1:3 soymilk-sized papers are larger than the other concentrations, demonstrating higher hydrophobic properties and water resistance. Restorers and painters often prefer using these concentrations of soymilk-sized paper in practical applications. Paper with low surface energy allows for controlled surface properties, such as producing a hydrophobic and self-cleaning coating. In Figure 2B, it can be observed that Shu Xuan paper, which exhibits the maximum contact angle, possesses the lowest surface energy. The smaller the water contact angle on the surface of the paper, the stronger the interaction between the water and the paper surface, indicating a higher surface energy. Therefore, Sheng Xuan paper, with the smallest water contact angle, has the largest surface energy. Regarding the different concentrations of soymilk-sized paper, the surface energy decreases with the increase in soymilk concentration. Previous studies have reported that the hydrophobic effect plays a significant role in protein structure stabilization and folding in soymilk [16]. The apolar part of soy protein establishes intramolecular contacts and releases water, while hydrogen bonds in the protein backbone contribute to the association and stabilization of the protein structure [17]. Among all the papers, the 1:1-sized sample, with the highest concentration of soymilk, exhibits an excellent folding performance of soy protein on the surface.

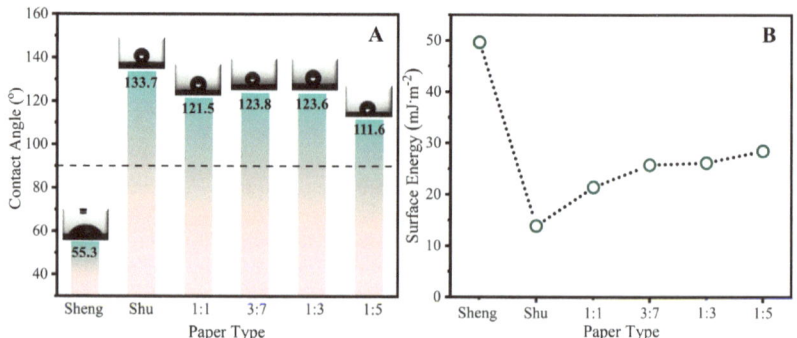

Figure 2. (**A**) Static contact angle and (**B**) surface energy of different papers.

Figure 3 presents scanning electron microscopy (SEM) images of Sheng Xuan paper, both in top view and side view, along with the corresponding elemental mapping of soymilk-sized Xuan paper. It is evident from the images that a membrane-like coating is observable only on the soymilk-sized paper. Upon further magnification, a larger number of well-ordered papillae with an average diameter of 75 nm can be observed on the surface.

The size distribution of papillae is mainly below 100 nm (inserted in Figure 3D), which is lower than the dynamic light scattering (DLS) result of soymilk, as shown in Figure 1C. Generally, the DLS technique provides the structure size of the aggregated state in liquid environments, which is larger than the SEM result. Conversely, the SEM images provide additional information about particle morphology and arrangement. Jiang's work has previously illustrated that when a large volume of water droplets is placed on lotus leaf surface, it tends to flow out more easily through the margins of these papillae, rather than overflowing onto the upper surface [18]. When the water contact angle is high, it indicates that water has difficulty spreading across the paper's surface, signifying lower water adhesion to the surface. As discussed above, the microstructured surface leads to a specific adhesion behavior, creating a high-energy barrier and contributing to the hydrophobic nature of the surface, with lower intrinsic surface energy. Furthermore, SEM-EDX analysis was employed to investigate the distribution of soy proteins within the paper. The SEM-derived elemental maps reveal an even distribution of C, N, O, and S elements, primarily originating from the soy protein. This observation further confirms that the soymilk permeates homogeneously into the paper fiber, ensuring a uniform distribution of soy protein throughout the paper structure.

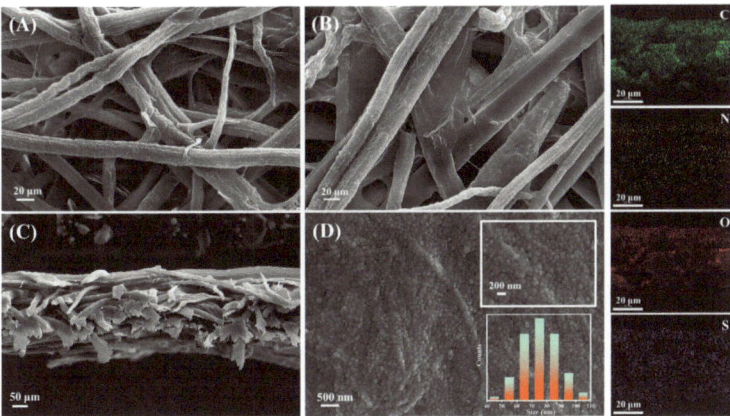

Figure 3. SEM images of (**A**) Sheng Xuan; soymilk-coated Xuan paper from the top (**B**) and side (**C**) view; and (**D**) the microstructure of soymilk-sized paper and corresponding elemental mapping.

The main component of paper is cellulose, which is a linear polymer consisting of linked D-glucopyranose units through β-1,4-glucosidic bonds. From a chain conformation perspective, cellulose can be viewed as an isotactic polymer of cellobiose [19]. The molecular structure of cellulose is depicted in Figure 4A. To investigate the structure of cellulose and its interaction with other molecules, nuclear magnetic resonance (NMR) techniques can be employed. Figure 4B,C present the ^{1}H and ^{13}C cross-polarization magic-angle spinning (CP/MAS) solid-state NMR spectra of Sheng Xuan and soymilk-sized paper. In the spectra, the chemical shift of 1.28 ppm can be attributed to protons in alkanes, while the peak at 29.55 ppm corresponds to methylene groups [20]. In addition, the proton peak at 4.5 ppm represents the hydroxyl proton at the C_3 position of cellobiose, as well as proton peaks from the glucopyranose backbone and the amino proton. In the soymilk-sized paper, new peaks appear. Two peaks at 2.01 and 2.23 ppm are characteristic of acetyl methyl proton derived from soymilk, as indicated by the dotted line in Figure 4B. The methyl proton peak from soymilk is observed at 0.89 ppm [21]. These additional peaks confirm the presence of soymilk components in the sized paper. CP/MAS NMR is a valuable technique for analyzing hydrogen bonds in solid cellulose. In the ^{1}H NMR spectra, a high-frequency shift at 26.28 ppm is observed in the soymilk-sized paper. The position of the hydrogen bond peaks in NMR spectra typically range from 8 to 20 ppm, while the peak at

26.28 ppm is significantly higher. However, previous studies have shown that the higher the strength of the hydrogen bond, the greater the peak shifts. This indicates the presence of stronger hydrogen bonds in the sized paper. In the ^{13}C NMR spectra (Figure 4C), the C_1 carbon peak of cellulose in Sheng Xuan paper appears at 104.8 ppm. In contrast, the C_1 carbon peak in soymilk-sized paper shifts to a higher magnetic field, indicating the presence of intramolecular hydrogen bonds. The peaks assigned to $C_{2,3,5}$ (70~80 ppm), C_4 (80~90 ppm), and C_6 (63.7 ppm) carbons remain unchanged [22]. The results obtained from the NMR spectra demonstrate a strong hydrogen-bonding interaction between soy protein and paper fibers. When a water-soluble protein undergoes folding into a compact and active conformation, the loss of polypeptide chain entropy is counteracted by favorable interactions within the protein and between the protein and its solvent. Non-covalent forces such as the hydrophobic effect, and hydrogen bonds, play significant roles in stabilizing the fold of the protein [16]. The soymilk-sized paper exhibits hydrophobicity and the presence of detected hydrogen bonds, both of which are essential for protein stability. These effects encourage the soluble protein to fold and remain in a stable state, which is supported by the observed surface energy characteristics.

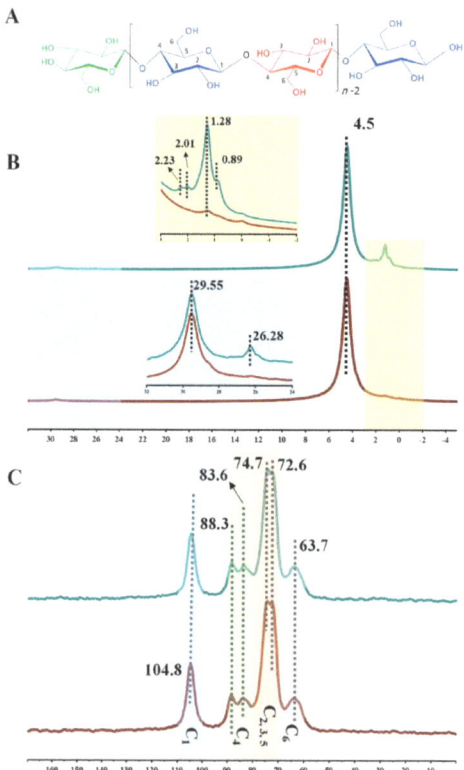

Figure 4. (**A**) The molecular structure of cellulose; (**B**) ^{1}H MAS NMR spectra and (**C**) ^{13}C CP/MAS spectra of Sheng Xuan paper (red, lower) and soymilk-sized paper (green, upper).

2.3. Paper Properties during Accelerated Aging

To investigate the long-term behavior of soymilk-sized paper, paper samples were subjected to dry-heat aging conditions at the temperature of 105 °C. The presence of acids in the paper can catalyze the hydrolytic degradation of cellulose, which is the inherent instability of paper. Over time, the paper undergoes acidification due to natural aging processes. Consequently, the degree of polymerization decreases, resulting in a loss of paper strength. As shown in Figure 5A, all the soymilk-sized paper samples exhibit an alkaline pH initially, which is favorable for conservators. When nearly neutral soymilk is applied to Xuan paper, the paper's fibers interact with the soymilk, leading to an initial increase in pH as the paper absorbs and interacts with the soymilk. However, the paper cellulose is ageing with the addition of soymilk. As the aging time progresses, the paper becomes progressively more acidic, resulting in a decrease in pH values. The most significant decline in pH occurs during the early stages of aging. The decrease in pH value is slower for paper samples with lower concentrations of soymilk. It is worth mentioning that the pH of soymilk-sized paper decreases at a much slower rate compared with Shu Xuan paper. The variation in pH among the soymilk-sized paper is influenced by the degradation and oxidation of the soy protein as well as the degradation of the paper cellulose. During the aging process, carboxylic acids, reactive oxygen species, and other products are generated from soymilk, promoting paper degradation. Hence, the pH values gradually decrease as the degradation progresses. Higher concentrations of soymilk result in faster degradation rates. As for the mechanical properties analysis of these various papers, it focuses on the folding endurance. In Figure 5B, the folding endurance of soymilk-sized paper is higher than that of Shu Xuan paper, and the folding resistance of all paper samples decrease with the prolongation of aging time. The trend is consistent with the observed pH values. The mechanical properties of the soymilk-sized paper exhibit improvements compared with the original Sheng Xuan paper, particularly during the aging process. Overall, the aging of soymilk-sized paper leads to acidification and a decline in pH values. However, the pH decrease is slower compared with traditional Shu Xuan paper. In addition, the soymilk-sized paper exhibits improved mechanical properties, particularly in terms of folding endurance, compared with the original Sheng Xuan paper.

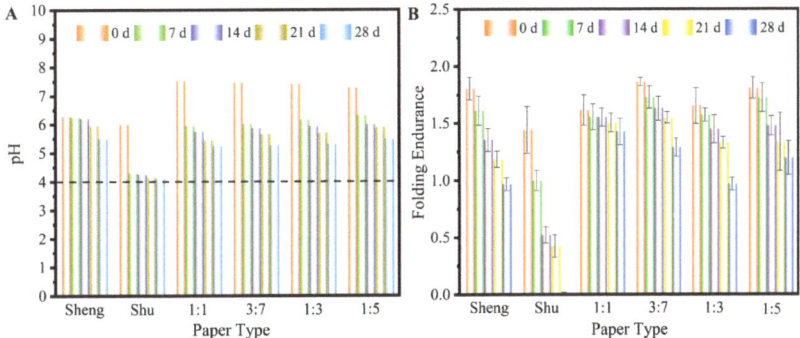

Figure 5. (**A**) The pH values and (**B**) folding endurance of different papers at various time stages.

Ms Qiu, the restorer, has observed that soymilk-sized paper exhibits a unique color that cannot be found in new paper. She specifically mentioned that the specified concentration of soymilk for sizing is one to three [6]. Figure 6 illustrates the chromaticity of different concentrations of soymilk-sized paper compared with the original Sheng Xuan paper. For all original paper samples, we compared the chromaticity changes with those of the Sheng Xuan paper. The results reveal that the 1:3 concentration of soymilk-sized paper exhibits the least chromaticity compared with the original Sheng Xuan paper. It is followed by 3:7, 1:1, and 1:5 soymilk-sized paper, and finally, the Shu Xuan paper. This finding suggests that the proposed concentration of 1:3 is reasonable and effective in replicating the color of historical paper artifacts during the restoration process. Furthermore, it is observed that the chromaticity of all the paper samples increases with the extension of aging time when compared with the corresponding original paper. The yellowing of cellulose during aging can be attributed to the presence of cellulose chromophores. The hydroxyl groups in the anhydroglucose units undergo transformations into carbonyl and carboxylic groups, resulting in the formation of aldehyde and ketone groups at positions C_2 and C_3. These changes contribute to the yellowing phenomenon of the paper [23]. Although soymilk sizing offers certain benefits, long-term preservation and careful monitoring are necessary to ensure the continued stability and longevity of the paper.

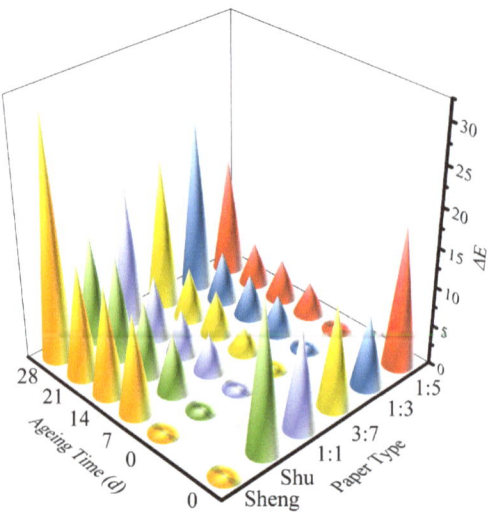

Figure 6. The chromaticity of different papers at various time stages.

2.4. Lifespan Expectancy of Different Paper

To better maintain and preserve paper, the Ekenstam equation was utilized to predict the lifespan of these paper samples. The initial DP values, final DP values, and degradation rates at ambient temperature were required to calculate the lifespan values. Previous studies have shown that paper strength reaches zero around a DP value of 200 [24]. In this study, the final DP value was set to 200, and some empirical equations were employed, as described in the Section 3. Figure 7A shows the degradation rate at an aging temperature of 105 °C. The rate constants were determined from the slopes of linear fitting, based on Equation (6). The rate constants for Sheng Xuan, Shu Xuan, and 3:7, 1:3, and 1:5 soymilk-sized Xuan paper were found to be 4.79×10^{-5}, 1.41×10^{-4}, 9.42×10^{-5}, 7.63×10^{-5}, and 6.10×10^{-5} d^{-1}, respectively. Subsequently, the degradation rate at ambient temperature was calculated using Equation (7). The pre-exponential factor in the equation is determined solely by the nature of paper degradation, independently of the reaction temperature and reactant concentration. Hence, the degradation constants at ambient temperature were calculated as 5.59×10^{-9}, 1.64×10^{-8}, 1.10×10^{-8}, 8.92×10^{-9}, and 7.21×10^{-9} d^{-1} for Sheng Xuan, Shu Xuan, and 3:7, 1:3, and 1:5 soymilk-sized Xuan paper, respectively. By reintroducing the final DP value of 200 into Equation (6), the lifespan of the paper samples was calculated and is depicted in Figure 7B. The results reveal that Sheng Xuan paper has the longest lifespan of 1385 years, while Shu Xuan paper has the shortest estimated lifespan of 396 years. This finding is consistent with previous conclusions, as the presence of alum in Shu Xuan paper promotes acid hydrolysis and facilitates paper degradation, creating an acidic environment within the pH range of 4.2 to 4.8, as observed in the pH characterization above [25]. For the different concentrations of soymilk-sized paper samples, the lifespan is negatively correlated with the concentration of soymilk. The longest estimated lifespan is 1153 years for 1:5 soymilk-sized paper, followed by 1:3 soymilk-sized paper with a lifespan of 987 years, and for 3:7 soymilk-sized paper, a lifespan of 680 years. However, an interesting phenomenon emerges where the folding endurance of 1:1 soymilk-sized paper surpasses that of 1:5 soymilk-sized paper after a month of accelerated aging. While soymilk may improve short-term properties like folding endurance, it may not provide sufficient protection against long-term degradation. The variation in lifespan among the soymilk-sized paper depends on the degradation and oxidation of the soy protein as well as the degradation of the paper cellulose. During the aging process, carboxylic acids, reactive oxygen species and other products are generated, promoting paper cellulose degradation [26–28]. In such cases, where the paper begins with an alkaline pH but gradually shift towards acidity over time, its alkaline reserve can be depleted. This depletion makes the paper more susceptible to the effects of aging. Higher concentrations of soymilk result in faster degradation rates. Nevertheless, it is important to note that these predictions provide a rough estimation of the usable life of the paper.

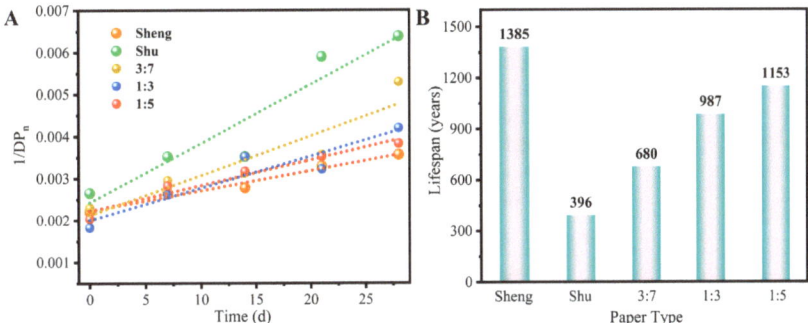

Figure 7. (**A**) Rate constants of different papers under accelerated aging conditions; (**B**) lifespan of different papers calculated using Ekenstam equation.

3. Materials and Methods

3.1. Materials

Commercially sourced soybeans were procured for this study. Soymilk extraction was carried out by a high-speed blender (WPB09J8, Westinghouse, Pittsburgh, PA, USA). Specifically, 600 mL water was added to 30 g soybeans to serve as an extraction solvent. The mixture was then processed in the high-speed blender for a duration of 6 min at room temperature. Subsequently, the raw soymilk was filtered through a cotton filter to eliminate the majority of insoluble impurities present in the soymilk. The resulting solution was then employed for further applications and subsequent analyses.

Diiodomethane was acquired from Shanghai Aladdin Biochemical Technology Co., Ltd., Shanghai, China, while Bis(ethylenediamine)copper(II) hydroxide (1 M in H_2O, or cupri-ethylenediamine) was supplied by Energy Chemical, Anhui Zesheng Technology Co., Ltd., Hefei, China, All chemicals were used as received without requiring additional purification. Furthermore, all solutions utilized throughout the experiment were prepared with deionized water. The initial handmade Xuan paper samples employed in this study were purchased from Hongxing Xuan Paper Co., Ltd. (Xuancheng, China). Specifically, writing-grade handmade Xuan paper composed of bark fiber from *Pteroceltis tatarinowii* and straw fiber was utilized. It is worth noting that Xuan paper can be categorized into various styles based on the extent of sizing. In this study, Sheng Xuan (unsized) paper and Shu Xuan (sized with alum–gelatin solution) paper were utilized as controls.

3.2. pH, Viscosity, Particle Size, and Zeta Potential of Soymilk

The stock soymilk was prepared following the procedures outlined in Section 3.1 above. To obtain different concentrations of soymilk, various volume ratios of the stock soymilk and deionized water were combined in 10 mL centrifuge tubes by pipettes. The pH values of the different soymilk concentrations were measured with a pH meter (FE28, Mettler–Toledo International Inc., Shanghai, China). Viscosity measurements of the different soymilk concentrations were conducted at a temperature of 20 °C using a Digital

Display Rotary Viscometer (NDJ-8S, Shanghai Lichen Bangxi Instrument Technology Co., Ltd., Shanghai, China).

The viscometer was equipped with different types of concentric cylinder geometry rotors. The surface charge and particle size of soymilk at various pH values were determined via a Zetasizer Nano ZS90 Analyzer. These measurements were conducted in triplicate to ensure accuracy. The pH values were adjusted within the range of 3.64 to 8.54 using a tiny amount of HNO_3 or $NaOH$ solutions. All measurements were performed at ambient temperature.

3.3. Basic Weight Measurement

To facilitate convenient sizing degree quantification, a paper cutter was employed to obtain circular samples with a diameter of 125 mm. Subsequently, the Sheng Xuan paper was immersed in various concentrations of soymilks to produce different types of soymilk-sized Xuan paper. After the sizing treatment, the paper was dried and stored under constant humidity and temperature conditions (humidity of 60 ± 2%, 25 ± 1 °C). The basic weight of the paper was calculated following the ISO 534:2011 standard [29]. In essence, the basic weight of the paper is determined by dividing the weight of the paper by its surface area.

3.4. Static Contact Angles and Surface Energy of Different Types of Paper

The static contact angles of water, observed from a side view, were measured with the OCA-20 contact-angle system (Dataphysics Instruments GmbH, Filderstadt, Germany) at room temperature. To determine the surface energy of the soymilk-sized paper, Owens theory was applied, which involves calculating the surface energy by two components: apolar and polar [30]. By measuring the contact angles of water and diiodomethane on the paper surface, the apolar and polar values of the paper's surface energy were obtained. The total surface energy of the paper is approximately the sum of these two components. It is known that the total surface tensions of water and diiodomethane are 72.8 and 50.8 mJ/m^2, respectively. The corresponding apolar parts are 21.8 and 48.5 mJ/m^2, while the polar parts are 51 and 2.3 mJ/m^2, respectively. By substituting the data of the surface tension and contact angle of water and diiodomethane into the equation below, two dependent equations can be derived. These equations allow for the calculation of the surface energy of the soymilk-sized paper.

$$\gamma_L(1+\cos\theta) = 2(\gamma_S^a \cdot \gamma_L^a)^{1/2} + 2\left(\gamma_S^p \cdot \gamma_L^p\right)^{1/2} \tag{1}$$

Herein, γ_L, γ_L^a, and γ_L^p represent the total, apolar, and polar surface tension of detecting liquid, respectively. The parameter θ is the static contact angle of the specific detecting liquid on the paper. γ_S^a and γ_S^p represent the apolar and polar parts of the solid surface energy that need to be determined.

3.5. Morphology Analysis of Different Types of Paper

A small quantity of each sample was affixed to conductive adhesive, enabling the investigation of the morphologies and distribution of elements on both the front and side surfaces of the paper using a HITACHI SU8010 scanning electron microscope (Hitachi High-Technologies corporation, Tokyo, Japan). Since the sample is non-conductive, it was necessary to apply a thin layer of gold onto the surface of the sample. This gold coating, consisting of several to a dozen atomic layers, is only a few nanometers to a dozen nanometers thick. Importantly, this gold coating has minimal impact on the appearance of the sample. Micrographs at various magnifications were captured at the energy of 10 KeV.

3.6. Solid-State Nuclear Magnetic Resonance Analysis

Due to the paper samples' insolubility in the typical solvents used for NMR analysis, the assessment of the paper's chemical structure was conducted using solid-state nuclear magnetic resonance (ssNMR). The spectra were acquired using a Bruker Avance II 400 MHz spectrometer (Bruker BioSpin GmbH, Ettlingen, Germany) operating at 400.18 MHz (^{1}H) and 100.62 MHz (^{13}C). Powdered samples were placed in a 4 mm rotor and spun at a rate of 10 kHz using a double air-bearing probe head. To obtain reliable results, each sample was subjected to 128 scans during the ssNMR analysis. The ^{13}C chemical shifts were referenced to adamantane, which serves as a standard reference compound in ssNMR spectroscopy.

3.7. Color Analysis of Different Types of Paper

The chromaticity of the paper was determined with an automatic colorimeter (ZB-A, Hangzhou Zhibang Automation Technology Co., Ltd., Hangzhou, China). To ensure accuracy and minimize potential errors, each sample was measured six times at different locations. The change in chromaticity ΔE was calculated via the following equation:

$$\Delta E = \sqrt{\Delta L^2 + \Delta a^2 + \Delta b^2} \qquad (2)$$

In the equation, L, a, and b represent three different colorimetric coordinates values, respectively. ΔL, Δa, and Δb are the differences between corresponding values of different samples.

3.8. pH Measurement and Mechanical Performance Testing

The measurement of paper pH differed from that of soymilk. In this study, a pH meter (pH-100B) provided by Shanghai Lichen Instrument Technology Co., Ltd. (Shanghai, China) was used. The measurement procedure followed the ISO 6588-1:2021 standard [31]. A drop of water was put on the paper surface, and the electrode head was vertically immersed in the water droplet. The pH value was recorded after allowing a 5 min equilibrium time. Each sample was tested five times at different locations, and the measurements were conducted at room temperature.

The mechanical performance of the paper was assessed by the folding endurance index, following the ISO 5626:1993 standard [32]. The paper specimen was subjected to longitudinal tension and folded forward and backward until it broke. The folding resistance was determined as the logarithm of the number of double folds before the specimen fractured. In this context, double folding refers to reciprocating the sample back and forth along the same fold line. The folding resistance is expressed as the logarithm of the number of double folds until the sample fractures. The MIT folding method was employed, and an MIT Folding Endurance Tester (Model: ZB-NZ135A) supplied by Hangzhou Zhibang Automation Technology Co., Ltd. (Hangzhou, China) was used for the folding endurance test.

3.9. Lifespan Evaluation of Different Types of Paper

The degree of polymerization (DP) of cellulose plays a crucial role in determining its physical and chemical properties, making it an important factor to consider when working with paper. The DP values of paper cellulose were measured following the ISO 5351:2010 standard [33]. The measurement procedure involved weighing a specific amount of paper and adding it to a 50 mL centrifuge tube along with 10 mL of deionized water. After allowing the paper fibers to swell, a 10 mL cupri-ethylenediamine solution was added, and the paper specimen was completely dissolved. The resulting solution was then transferred to an Ubbelodhe viscometer, and the specific viscosity (η_{sp}) of the dissolved solution relative to the cupri-ethylenediamine solvent was recorded. Importantly, the temperature was maintained at 25 °C throughout the measurement. After obtaining the specific viscosity, the intrinsic viscosity ($[\eta]$) was calculated via the Martin empirical equation below [34]:

$$\eta_{sp} = [\eta] \cdot \rho \cdot e^{K[\eta]\rho} \qquad (3)$$

Herein, ρ represents the concentration of paper cellulose (in g/mL), and K is an empirical constant specific to the cellulose–copper ethylenediamine system, equal to 0.13.

Subsequently, the molecular weight (M) of cellulose, which is usually expressed as the DP, was determined via the Mark–Houwink equation:

$$[\eta] = KM^{\alpha} \qquad (4)$$

In this equation, K and α are constants specific to the system. Hence, the Mark–Houwink equation can be rewritten as follows:

$$DP^{0.905} = 0.75[\eta] \qquad (5)$$

The paper's life expectancy can be predicted via the Ekenstam equation:

$$\frac{1}{DP_n} - \frac{1}{DP_0} = kt \qquad (6)$$

where DP_n and DP_0 are the paper's degree of polymerization at any given moment and the initial moment, respectively. k (day^{-1}) is the rate constant of degradation, and t is the accelerated aging time of the paper sample. Since the accelerated aging experiment was conducted at 105 °C, the obtained degradation rate constant should be converted to the corresponding rate constant at 25 °C. This conversion can be accomplished via the Arrhenius equation:

$$\ln k = \ln A - \frac{E_a}{RT} \qquad (7)$$

where A represents the pre-exponential factor, which is temperature-independent. E_a denotes the activation energy, which has been experimentally determined to be 106 kJ/mol. R is the gas constant, and T is the absolute temperature [35]. By obtaining the degradation rate constant, defining the DP limit as 200, and plugging the relevant values into the Ekenstam equation, the lifespan of the paper can be evaluated [36].

4. Conclusions

This study focused on the application of soymilk as a sizing agent for Xuan paper and its effects on various properties and long-term behaviors, especially for the potential mechanism of sizing paper with soymilk. The soymilk-sized paper exhibited hydrophobicity, improved mechanical properties, and a unique chromaticity that was not found in new paper. The microstructure, protein folding, and hydrogen-bonding interactions between soy protein and paper fibers contributed to these properties. Moreover, the soymilk-sized paper showed resistance to acidification, a slower pH decrease, and decreased basic weight compared with traditional alum–gelatin-solution-coated paper. The aging process led to acidification, decreased mechanical strength, increased chromaticity, and degradation of the paper samples. Predictions using the Ekenstam equation estimated the lifespan of the paper, with Sheng Xuan paper having the longest estimated lifespan and soymilk-sized paper exhibiting negative correlations between lifespan and soymilk concentration. These findings provide valuable insights for the preservation and long-term maintenance of paper using soymilk sizing, in spite of presenting a rough estimation, and further preservation strategies should be considered for optimal paper conservation.

Author Contributions: Conceptualization, H.Z. (Hongdong Zhang); Material preparation, data collection and analysis, C.W., Y.L. (Yangyang Liu), Y.H. and M.D.; Writing—original draft preparation, C.W.; Writing—review and editing, Y.L. (Yixin Liu), Y.H., X.C., Y.L. (Yangyang Liu), P.L., H.Z. (Hongbin Zhang) and Y.Y.; Funding acquisition, H.Z. (Hongbin Zhang). All authors have read and agreed to the published version of the manuscript.

Funding: This work was supported by the National Natural Science Foundation of China (22175040) and the Foundation of State Key Laboratory of Biobased Material and Green Papermaking, Qilu University of Technology, Shandong Academy of Sciences (GZKF202109, GZKF202210).

Institutional Review Board Statement: Not applicable.

Informed Consent Statement: Not applicable.

Data Availability Statement: The authors confirm that the data supporting the findings of this study are available within the article.

Conflicts of Interest: The authors declare that they have no known competing financial interests or personal relationships that could have appeared to influence the work reported in this paper.

References

1. Administration of Cultural Heritage. Data Bulletin of the First National Census of Movable Cultural Relics. Available online: http://www.ncha.gov.cn/art/2017/4/7/art_1984_139587.html (accessed on 5 July 2023).
2. Jiang, X. The Application of "Repairing the Old as the Old" in Painting and Calligraphy Restoration. Master's Thesis, Nanjing University of the Arts, Nanjing, China, 2019.
3. Lee, H.; Kim, H.S. Pasting and paste properties of waxy rice starch as affected by hydroxypropyl methylcellulose and its viscosity. *Int. J. Biol. Macromol.* **2020**, *153*, 1202–1210. [CrossRef] [PubMed]
4. Xu, K.; Wang, J. Discovering the effect of alum on UV photo-degradation of gelatin binder via FTIR, XPS and DFT calculation. *Microchem. J.* **2019**, *149*, 103934. [CrossRef]
5. Taniguchi, M.; Saito, K.; Nomoto, T.; Namae, T.; Ochiai, A.; Saitoh, E.; Tanaka, T. Identification and characterization of multifunctional cationic and amphipathic peptides from soybean proteins. *Biopolymers* **2017**, *108*, e23023. [CrossRef]
6. Qiu, J. The Application of Soymilk in the Restoration of Painting and Calligraphy. Available online: http://art.people.com.cn/n1/2016/1017/c206244-28785176.html (accessed on 5 July 2023).
7. Wu, S. *Basic Techniques of Freehand Figure Painting*, 1st ed.; Hangzhou Academy of Fine Arts Press: Hangzhou, China, 1988; p. 36.
8. Xu, W.; Zhu, P. Application of soy milk in Chinese painting and paper conservation. *Sci. Conserv. Archaeol.* **2012**, *24*, 1–4.
9. Wu, H.; Qi, B.; Jiang, L.; Li, Y.; Feng, H.; Cao, L.; Ma, W.; Ding, J.; Wang, R. Effect of thermal properties and spatial conformation of soybean protein isolate on surface hydrophobicity. *J. Chin. Cereals Oils Assoc.* **2014**, *29*, 42–46.
10. He, Q.; Wang, L.; Xu, K.; Wang, J. Application of soybean water with different heating temperatures to Xuan paper. *Sci. Conserv. Archaeol.* **2019**, *31*, 8–13.
11. Vnučec, D.; Kutnar, A.; Goršek, A. Soy-based adhesives for wood-bonding—A review. *J. Adhes. Sci. Technol.* **2016**, *31*, 910–931. [CrossRef]
12. Li, D.; Li, X.; Wu, G.; Li, P.; Zhang, H.; Qi, X.; Wang, L.; Qian, H. The characterization and stability of the soy protein isolate/1-Octacosanol nanocomplex. *Food Chem.* **2019**, *297*, 124766. [CrossRef]
13. Peng, Y.; Kersten, N.; Kyriakopoulou, K.; van der Goot, A.J. Functional properties of mildly fractionated soy protein as influenced by the processing pH. *J. Food Eng.* **2020**, *275*, 109875. [CrossRef]
14. O'Flynn, T.D.; Hogan, S.A.; Daly, D.F.M.; O'Mahony, J.A.; McCarthy, N.A. Rheological and solubility properties of soy protein isolate. *Molecules* **2021**, *26*, 3015. [CrossRef]
15. Liu, X.; Hsieh, Y.L. Amphiphilic protein microfibrils from ice-templated self-assembly and disassembly of Pickering emulsions. *ACS Appl. Bio. Mater.* **2020**, *3*, 2473–2481. [CrossRef]
16. Fiedler, S.; Broecker, J.; Keller, S. Protein folding in membranes. *Cell. Mol. Life Sci.* **2010**, *67*, 1779–1798. [CrossRef]
17. Cserhati, T.; Szogyi, M. Interactions between proteins, peptides and amino acids. New advances 1986–1989. *Nahrung* **1990**, *34*, 803–810. [CrossRef] [PubMed]
18. Liu, M.; Jiang, L. Switchable adhesion on liquid/solid interfaces. *Adv. Funct. Mater.* **2010**, *20*, 3753–3764. [CrossRef]
19. Wertz, J.L.; Bédué, O.; Mercier, J.P. *Cellulose Science and Technology*, 1st ed.; EPFL Press: Lausanne, Switzerland, 2010; pp. 21–24.
20. Chierotti, M.R.; Gobetto, R. Solid-state NMR studies of weak interactions in supramolecular systems. *Chem. Commun.* **2008**, *14*, 1621–1634. [CrossRef]
21. Hinds, M.G.; Norton, R.S. NMR spectroscopy of peptides and proteins. *Mol. Biotechnol.* **1997**, *7*, 315–331. [CrossRef] [PubMed]
22. Kamide, K. *Cellulose and Cellulose Derivatives Molecular Characterization and Its Application*, 1st ed.; Elsevier Science: Amsterdam, The Netherlands, 2005; pp. 460–600.
23. Carter, H.A. The chemistry of paper preservation. Part 2. The yellowing of paper and conservation bleaching. *J. Chem. Educ.* **1996**, *73*, 1068–1073. [CrossRef]
24. Zou, X.; Uesaka, T.; Gurnagul, N. Prediction of paper permanence by accelerated aging I. Kinetic analysis of the aging process. *Cellulose* **1996**, *3*, 243–267. [CrossRef]
25. Carter, H.A. The chemistry of paper preservation. Part 4. Alkaline paper. *J. Chem. Educ.* **1997**, *74*, 508–511. [CrossRef]
26. Xia, M.; Chen, Y.; Guo, J.; Feng, X.; Yin, X.; Wang, L.; Wu, W.; Li, Z.; Sun, W.; Ma, J. Effects of oxidative modification on textural properties and gel structure of pork myofibrillar proteins. *Food Res. Int.* **2019**, *121*, 678–683. [CrossRef]
27. Schrader, E.K.; Harstad, K.G.; Matouschek, A. Targeting proteins for degradation. *Nat. Chem. Biol.* **2009**, *5*, 815–822. [CrossRef] [PubMed]
28. Zhang, G.; Zhu, C.; Walayat, N.; Nawaz, A.; Ding, Y.; Liu, J. Recent development in evaluation methods, influencing factors and control measures for freeze denaturation of food protein. *Crit. Rev. Food Sci. Nutr.* **2022**, *63*, 5874–5889. [CrossRef] [PubMed]
29. Paper and Board–Determination of Thickness, Density and Specific Volume: ISO 534:2011. Available online: https://www.iso.org/standard/53060.html (accessed on 21 September 2023).

30. Chen, Z.; Nosonovsky, M. Revisiting lowest possible surface energy of a solid. *Surf. Topogr. Metrol. Prop.* **2017**, *5*, 045001. [CrossRef]
31. Paper, board and pulps–Determination of pH of aqueous extracts–Part 1: Cold extraction: ISO 6588–1:2021. Available online: https://www.iso.org/standard/83250.html (accessed on 21 September 2023).
32. Paper–Determination of Folding Endurance: ISO 5626:1993. Available online: https://www.iso.org/standard/11700.html (accessed on 21 September 2023).
33. Pulps–Determination of Limiting Viscosity Number in Cupri–Ethylenediamine (CED) Solution: ISO 5351:2010. Available online: https://www.iso.org/standard/51093.html (accessed on 21 September 2023).
34. Liu, P.; Zhang, H.; Wang, S.; Gao, B.; Yan, Y.; Tang, Y. Improvement on measuring the intrinsic viscosity of cellulose by the copper ethylenediamine method. *Daxue Huaxue* **2020**, *36*, 2010068. [CrossRef]
35. Jin, C.; Wu, C.; Liu, P.; Yu, H.; Yang, Y.; Zhang, H. Kinetics of cellulose degradation in bamboo paper. *Nord. Pulp Pap. Res. J.* **2022**, *37*, 480–488. [CrossRef]
36. Zhang, X.; Liu, P.; Yan, Y.; Yao, J.; Tang, Y.; Yang, Y. Degradation of Chinese handmade papers with different fiber raw materials on molecular and supramolecular structures. *Poly. Degrad. Stab.* **2023**, *211*, 110330. [CrossRef]

Disclaimer/Publisher's Note: The statements, opinions and data contained in all publications are solely those of the individual author(s) and contributor(s) and not of MDPI and/or the editor(s). MDPI and/or the editor(s) disclaim responsibility for any injury to people or property resulting from any ideas, methods, instructions or products referred to in the content.

Article

Study on the Properties of FEVE Modified with Ag$_2$O/OH-MWCNTS Nanocomposites for Use as Adhesives for Wooden Heritage Objects

Gele Teri [1], Cong Cheng [1], Kezhu Han [1], Dan Huang [1], Jing Li [2], Yujia Luo [1,*], Peng Fu [3,*] and Yuhu Li [1,*]

[1] Engineering Research Center of Historical Cultural Heritage Conservation, Ministry of Education, School of Materials Science and Engineering, Shaanxi Normal University, Xi'an 710119, China; terigelesnnu@163.com (G.T.); congcheng2017@snnu.edu.cn (C.C.); hankekezhu@126.com (K.H.); hdansnnu@163.com (D.H.)
[2] Shandong Museum, Jinan 250014, China; lijing9669@126.com
[3] Shaanxi Institute for the Preservation of Culture Heritage, Xi'an 710075, China
* Correspondence: yujialuo@snnu.edu.cn (Y.L.); fupeng@snnu.edu.cn (P.F.); liyuhu@snnu.edu.cn (Y.L.)

Citation: Teri, G.; Cheng, C.; Han, K.; Huang, D.; Li, J.; Luo, Y.; Fu, P.; Li, Y. Study on the Properties of FEVE Modified with Ag$_2$O/OH-MWCNTS Nanocomposites for Use as Adhesives for Wooden Heritage Objects. *Molecules* **2024**, *29*, 1365. https://doi.org/10.3390/molecules29061365

Academic Editors: Yueer Yan, Yi Tang and Yuliang Yang

Received: 19 February 2024
Revised: 14 March 2024
Accepted: 15 March 2024
Published: 19 March 2024

Copyright: © 2024 by the authors. Licensee MDPI, Basel, Switzerland. This article is an open access article distributed under the terms and conditions of the Creative Commons Attribution (CC BY) license (https://creativecommons.org/licenses/by/4.0/).

Abstract: The durability of wooden heritage objects and sites can be affected by external environmental factors, leading to decay, cracking, and other forms of deterioration, which might ultimately result in significant and irreversible loss. In this study, a FEVE resin was modified with Ag$_2$O/OH-MWCNTS (MA), denoted as MAF, where three concentrations were prepared using in situ precipitation, and the resulting composite adhesive was characterized by a high viscosity and effective bacteriostatic properties, demonstrating a better viscosity and thermal stability, as well as antibacterial properties, than pure FEVE resin. The results show that MAF adhesives present good thermal stability, as evidenced by a lower mass loss rate following treatment at 800 °C compared to the pure FEVE resin. At a consistent shear rate, the viscosity of MAF demonstrates a notable increase with the proportion of MA, which is better than that of FEVE. This suggests that the nano-Ag$_2$O particles in MA act as physical crosslinking agents in FEVE, improving the viscosity of the composite adhesive MAF. The adhesion strength between MAF and wood exhibits a similar trend, with wooden samples showing higher shear strengths as the proportion of MA increases in comparison to FEVE. Simultaneously, the antibacterial effects of the MAF adhesive exceeded 1 mm for *Trichoderma*, *Aspergillus niger*, and *white rot fungi*. The antibacterial activity of the MAF adhesive exhibited a direct correlation with the concentration of Ag$_2$O/OH-MWCNTS, with the most pronounced inhibitory effect observed on Trichoderma. The MAF adhesive demonstrates promising prospects as an adhesive for wooden heritage artifacts, offering a novel approach for the rapid, environmentally friendly, and efficient development of composite adhesives with superior adhesive properties.

Keywords: Ag$_2$O/OH-MWCNTS-FEVE; adhesives for wooden artifacts; bonding strength; antibacterial

1. Introduction

Wood, as a readily accessible and easily processed natural resource, has been utilized throughout history in various aspects of human life. In ancient civilizations, when alternative materials were scarce, wood was extensively employed for the construction and ornamentation of historical sites [1,2]. Across different cultures and time periods, a plethora of intricate wooden artifacts, crafts, and structures have been preserved as cultural relics, frequently documented in historical records [3]. During the middle Paleolithic Age (~400,000 years ago) various heritage artifacts and sites, such as sculptures, lacquered wooden ware, bridges, defense weapons, wooden buildings of the Imperial Palace, carriages, ships, and other items, were created [4,5]. These wooden cultural heritage items serve as valuable materials for research on ancient history, culture, and living standards, and therefore possess significant historical and artistic value [4,6–8]. Furthermore, these artifacts and sites, which cannot be replicated, exemplify the ingenuity and resourcefulness of humanity.

Wood is a natural bio-composite whose physical and chemical properties change as a result of various factors during the preservation process, leading to varying degrees of degradation. The loss of mechanical stability due to degradation significantly impacts the lifetime and structure of wooden cultural relics [9–14]. Cellulose, hemicellulose, and lignin are the primary components of wood, where the free hydroxyl groups are abundant. The fluctuation of temperature and humidity in the preservation environment allows for the free absorption and release of water vapor by the hydroxyl groups in wood, leading to changes in the wood's humidity. The synergistic effect caused by these factors can result in alterations to the size of the wood and the development of cracks [15–17], which might ultimately result in the deterioration and loss of wooden cultural relics, leading to significant and irreparable losses in historical research and cultural heritage. Hence, the preservation of wooden cultural artifacts is imperative and significant, necessitating the advancement of appropriate adhesives for such objects [18].

According to the definition by Encyclopedia Britannica, an adhesive refers to any substance that is capable of holding materials together [19]. The history of adhesives is closely associated with human history, becoming one of the most extensively used materials in early human societies with the advent and evolution of composite tools during the Mesolithic era. Adhesives, initially employed for hafting, were found to have wide-ranging applications in painting, decoration, lacquerware, bows and arrows, construction, and pottery repair [20–25]. In ancient times, a variety of adhesive materials were utilized, including animal glues, eggs, casein, blood, plant resins, gums, starch, bitumen, tar, wax, and oils. With technological advancements, the diversity of adhesives has expanded, including various organic and inorganic materials. Common adhesives used in the conservation of wooden cultural heritage artifacts include natural adhesives such as casein, rabbit skin glue, fish glue, waxes, oils, and natural resins, as well as synthetic adhesives like the copolymers of methyl methacrylate and ethyl acrylate (Paraloid B72) and polyvinyl acetate emulsions (Ravemul M18—Vinavil). However, these adhesives have their respective advantages and disadvantages. For instance, animal glues are natural, non-toxic, fully reversible, and exhibit excellent adhesion to wooden surfaces without staining, but can be susceptible to microbial growth.

A lot of attention has been paid to FEVE fluorocarbon [26,27] resin in the field of coating research and development owing to its unique structural properties (Figure 1). Currently, research on FEVE primarily revolves around the development of coatings and the regulation of their performance. The high bond energy of the C-F fluorine-containing bond in fluorocarbon resin at ~485 kJ/mol is attributed to the large electronegativity of fluorine atoms [28]. This results in a polymer with significant repulsive forces between adjacent fluorine atoms, leading to a high stability and exceptional aging resistance. Additionally, FEVE can be cured and shaped at room temperature, showing a low surface energy, strong adhesion properties, and other favorable characteristics, making it a suitable option for wood adhesives. However, challenges such as low tensile strength and inadequate bacteriostatic properties still exist.

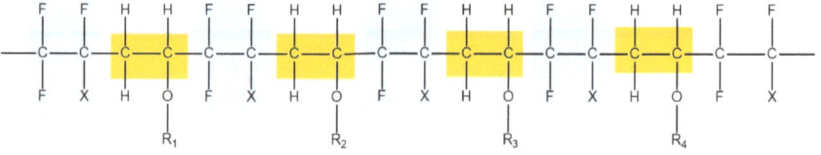

X: CF_3/F/Cl

R_1-R_4: Alkyl / Naphthenic Base / Hydroxyalkyl / Carboxyalkyl

Figure 1. Molecular formula of FEVE.

Wood is prone to microbial degradation, particularly by fungi, leading to a decrease in mechanical stability [29]. Microbial corrosion can affect both wood itself and adhesive materials, indicating the importance of antibacterial properties in protecting wooden cultural artifacts from wood-rot fungi [30,31]. In recent years, antibacterial nanomaterials have emerged as promising candidates for antimicrobial applications within the field of chemical antibacterial technologies, owing to their expansive specific surface area and distinctive chemical and physical characteristics [32,33]. Numerous nanomaterials, such as ZnO, CuO, TiO_2, and Ag_2O, have demonstrated exceptional antibacterial properties and are utilized as antibacterial agents [34–37]. Notably, silver oxide nanoparticles offer advantages such as ease of production, cost-effectiveness, and potent antibacterial and antifungal activity. Silver oxide antibacterial nanoparticles (NPs) are commonly utilized for their notable antibacterial efficacy against a diverse range of bacteria [38–40], leading to their utilization as a preservative in various medical devices, food packaging, and environmental purification procedures [41–43]. Nonetheless, nano Ag_2O is hindered by inadequate dispersion. Then, carbon nanotubes (CNTs) began to be used as a novel class of nanomaterials following their inception in 1991 [44], offering substantial application benefits across numerous scientific and technological domains [45–47]. Hydroxylated multi-walled carbon nanotubes (OH-MWCNTS) are a novel carbon-based material with superior properties such as increased weather resistance, environmental compatibility, corrosion resistance, lightweight construction, expansive surface area, excellent thermal stability, and robust biological compatibility when compared to similar materials. Currently, hydroxylated MWCNTS are extensively utilized in chemical reactions, environment protection, biological classification and conversion, electrochemical energy storage, and other related fields [48–51]. Important findings have been reported based on studies of MWCNTs/FEVE composite coatings [52]. MWCNTs can be uniformly distributed within the composite coating regardless of the amount. As the content of MWCNTs increases in the coating, both the glossiness and the static friction coefficient of the coating significantly decrease. Also, the surface of the coating becomes rougher, and the composite coating exhibits both hydrophobic and oleophobic properties. Adding a small amount of MWCNTs to the coating can greatly improve the conductivity of the MWCNTs/FEVE composite coating. The incorporation of hydroxyl groups may enhance the interaction between MWCNTs and FEVE resin, thereby improving the load transfer efficiency and interface bonding strength of the composite. The exceptional chemical properties and weathering resistance of FEVE resin make it ideal for the conservation of outdoor heritage objects. The addition of OH-MWCNTs can further enhance these properties, leading to a composite which is suitable for harsh environmental conditions. The introduction of OH-MWCNTs may also affect the surface properties of the composites, including roughness and hydrophobicity/hydrophilicity.

Based on the above discussion, this study involved the preparation of a stable and low-cost composite material consisting of Ag_2O-decorated hydroxyl multi-wall carbon nanotubes, i.e., Ag_2O/OH-MWCNTS (MA), using an in situ precipitation method. This composite material was then mixed with a FEVE resin to prepare a durable and bacteriostatic wooden adhesive, i.e., Ag_2O/OH-MWCNTS-FEVE (MAF), which was subsequently subjected to the exploration of variolous properties. The structural and performance characteristics of the resulting nanomaterials were analyzed using techniques such as infrared spectroscopy, X-ray diffraction, and SEM spectroscopy. The thermal stability and viscosity of the MAF wood adhesive were assessed, while the shear strength and antibacterial properties of the wood adhesive were investigated using a universal tensile testing machine and an antibacterial zone test. Comparative analysis revealed that the modified MAF exhibited superior viscosity and bacteriostatic properties compared to the original FEVE, suggesting a promising application potential for the composite material.

2. Results and Discussion

2.1. X-ray Diffraction Analysis

Each crystalline has a distinct atomic arrangement and exhibits a characteristic X-ray diffraction pattern, serving to identify its crystal structure. X-ray diffraction (XRD), a nondestructive method, is commonly employed to determine the crystal structure and purity of nanoparticles [53]. The XRD patterns of the nanocomposites Ag_2O/OH-MWCNTS and OH-MWCNTS are shown in Figure 2a. The 26.1o and 42.9o in the XRD patterns of OH-MWCNTS correspond to the reflections of the surfaces of hexagonal graphite (JPCDS No. 75-1621) (002) and (101). The diffraction angles (2θ) at 26.2°, 32.9°, 38.1°, 55.2°, 65.7°, and 69.1°, correspond to the groups of Ag_2O lattice planes (cubic structure) (110), (111), (200), (220), (311), and (222), respectively [JCPDS No.76-1393]. The presence of a prominent intensity peak in the (111) plane suggests a well-ordered arrangement of lattice atoms [54]. Furthermore, the absence of peaks corresponding to other substances in the X-ray diffraction pattern suggests a high concentration of Ag_2O within the nano-composite Ag_2O/OH-MWCNTS.

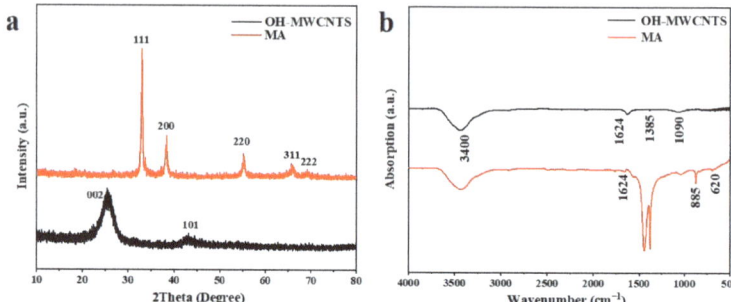

Figure 2. (a) XRD spectra and (b) FTIR spectra of the OH-MWCNTS and MA.

2.2. Infrared Absorption Spectroscopy Analysis

Fourier transform infrared spectroscopy (FTIR) spectroscopy was used to investigate the characteristics of the functional groups of Ag_2O/OH-MWCNTS and OH-MWCNTS as shown in Figure 2b. The broad band at 3400 cm^{-1} indicates the O–H stretching vibrations of the hydroxyl groups corresponding to H-bonded alcohols and also to intramolecular H bonds, which confirms the existence of the O-H group on the surface of the OH-MWCNTs [55]; a band at 1624 cm^{-1} is attributed to the stretching vibrations in the carbon backbones of OH-MWCNTs. Based on findings from previous studies, metal–oxygen stretching frequencies typically present within the range of 500 to 600 cm^{-1} [56]. In Figure 2b, the peak shown at 620 cm^{-1} is attributed to the Ag-O vibration of Ag_2O [57]. The absorption peak at 885 cm^{-1} corresponds to the Ag-O bond [58], while the peak at 1271 cm^{-1} is notably strong, indicating a high Ag_2O content in Ag_2O/OH-MWCNTS. These results are consistent with the XRD data, providing further evidence for the formation of Ag_2O/OH-MWCNTS.

2.3. SEM Analysis

Figure 3 presents scanning electron microscope (SEM) surface morphology images of the nano-composites (Ag_2O/OH-MWCNTS) at various magnifications. The images reveal a uniform near-spherical nanobeam of Ag_2O with a diameter of 19–59 nm, with OH-MWCNT dispersed among the nanobeams, which is consistent with findings from FTIR and XRD analyses. Typically, smaller particle sizes are more beneficial for improving antibacterial activity, which is due to the reduced particle size; a greater number of particles can readily adhere to the bacterial cell membrane surface, facilitating successful cell penetration and subsequent destruction of the cell's physiological functional groups [59]. The SEM images indicate that the in situ synthesis of nano-composites (Ag_2O/OH-MWCNTS) offers a straightforward and efficient method to achieve the requisite size specifications for Ag_2O.

Figure 3. SEM images of the Ag_2O/OH-MWCNTS with different magnifications.

2.4. Thermal Stability Analysis

The thermal properties of the FEVE and MAF composites were examined using thermogravimetric (TG) analysis, and the results are presented in Figure 4a. The DTG curves exhibit a general similarity among the samples. The initial peaks observed in the TG curves of MAF-1, MAF-2, and MAF-3 between 150 and 220 °C are attributed to the evaporation of surface and interlayer moisture. Subsequently, a second peak is observed between 330 and 470 °C, indicating the degradation of oxygen-containing functional groups on the surface (OH-MWCNTS) and the degradation of chemical bonds within the FEVE resin. Given the low content of MWCNT [60], no weight loss induced by MWCNT oxidation was found at temperatures ranging from 600 to 800 °C. Mass losses of 76.7%, 72.1%, and 69.9% were observed for MAF-1, MAF-2, and MAF-3, respectively, in the temperature range of 200–800 °C. In comparison, the FEVE resin presents a mass loss of 82.5%, indicating a gradual decrease in mass loss with increasing concentrations of Ag_2O/OH-MWCNTS. The incorporation of Ag_2O/OH-MWCNTS resulted in the enhanced thermal stability of FEVE.

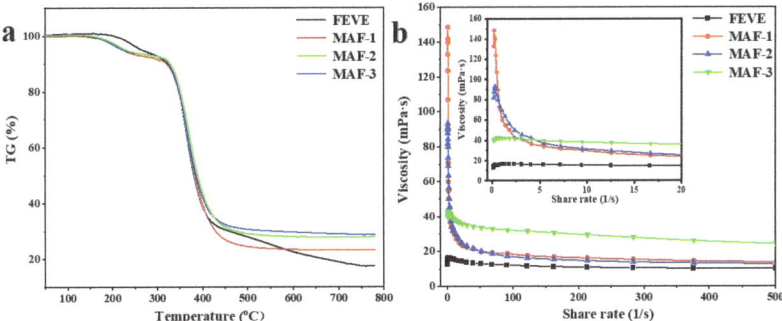

Figure 4. (**a**) TG spectra of the MAF adhesive with different concentrations; (**b**) viscosity–shear rate diagram of the MAF adhesive with different concentrations.

2.5. Viscoelastic Measurement

The viscosity of the adhesive plays a crucial role in determination of its permeability and hydrophilicity on the substrate surface, as well as its adhesion properties [61]. Figure 4b shows the change in apparent viscosity of the composite adhesive MAF with different concentrations of MA added as the shear rate increases. It is evident that the shear thinning effect of FEVE viscosity is relatively small when the shear rate increases. FEVE possesses a multitude of C-F chemical bonds, with the C-F bond energy reaching a remarkable

485.6 kJ/mol, resulting in minimal polarity and a stable molecular structure. From a structural perspective, the FEVE structure comprises three Fs, forming a spiral-like three-dimensional arrangement that tightly encircles each C-C bond within the molecule, filling in the gaps between C-C bonds, thereby ensuring maximum structural integrity and tightness [28]. Therefore, the viscosity of FEVE changes little with the increase in the shear rate. When the shear rate is constant, the apparent viscosity of MAF increases with the increase in the MA concentration. Upon investigation, nano-Ag_2O was immobilized on the surface of OH-MWCNTS, leading to an enhanced specific surface area of the material. Nano-sized Ag_2O particles play a role as physical crosslinking points in FEVE, thereby increasing the viscosity of the composite adhesive M [28,62].

MAF-1 and MAF-2 rapidly decrease in the low shear rate region, gradually decrease in the high shear rate region, and reach the viscosity of infinite shear rate. When the shear rate of MAF-1 and MAF-2 was increased from 0.1 to 20 s^{-1}, the viscosity decreased significantly, indicating that the composite adhesive exhibited shear thinning behavior and had pseudoplastic properties. This is due to the low content of MA, which results in fewer physical crosslinking points, leads to unstable intermolecular forces and intermolecular interactions that are easily disrupted under high shear rates. The content of Ag_2O/OH-MWCNTS in MAF-3 forms a sufficient number of physical cross-linking points, so the viscosity changes slightly with the increase in shear rate. Due to the effect of MA on the viscosity of the composite adhesive MAF, this will affect the adhesion between the adhesive and the substrate, so we verified its bonding strength.

2.6. Analysis of Adhesive Bonding Shear Strength

The cohesion and adhesion of MAF play crucial roles as the adhesive in composite nanomaterials. When the water-based FEVE interacts with the surface of wood, it exhibits permeability characteristics. And, with the addition of MA, the viscosity of the composite adhesive MAF increases, thus reducing the permeability of MAF on the surface of porous wooden materials. The binding force necessitates the formation of a continuous crystalline phase structure, with the strength of the bond being largely dependent on the molecular attraction between the adhesive and the substrate surface [63]. Figure 5 shows the shear strength and stress–strain curve changes of wooden panels using FEVE, MAF-1, MAF-2, and MAF-3 adhesives. The shear strength values for the wood fixed using FEVE, MAF-1, MAF-2, and MAF-3 were 0.994 MPa, 1.32 MPa, 1.42 MPa, and 1.73 MPa (Table 1), respectively. After 5 days of degradation, the bonded wooden samples exhibited a notable decrease in shear strength compared to the samples before degradation. The reductions in shear strength after degradation were measured at 51.71%, 45.45%, 30.99%, and 28.90% for wooden samples bonded with FEVE, MAF-1, MAF-2, and MAF-3, respectively. Although the shear strength of the FEVE adhesives decreased under high relative humidity, the incorporation of MA led to a notable increase in the adhesive strength of MAF. With the increase in MA, the shear strength of plywood increased, in both the dry and after high RH degradation. As the content of MA increases, the shear strength of the wooden panel increases. The results show that the addition of MA can improve the bonding properties between FEVE and the wooden panel. It is reported that carbon nanotubes added as fillers in fluorocarbon coatings can form functional networks and improve the compactness of the coating. Nano-sized Ag_2O is deposited on the surface of OH-MWCNTS, increasing the specific surface area of OH-MWCNTS, and the number of physical cross-linking points [64–66]. Small-size inorganic Ag_2O nanoparticles can effectively fill the structural micropores generated during the curing process. Therefore, the bonding force between MAF and the wooden panel is increased. The stress–displacement curve directly reflects the properties of the material itself. As the content of MA increases, the stress on the composite adhesive MAF increases. Under the same stress, the displacement of FEVE is greater than that of MAF-1, MAF-2, and MAF-3. The results showed that the addition of MA increased the brittleness of the composite adhesive MAF material, which was related to the formation of a dense coating.

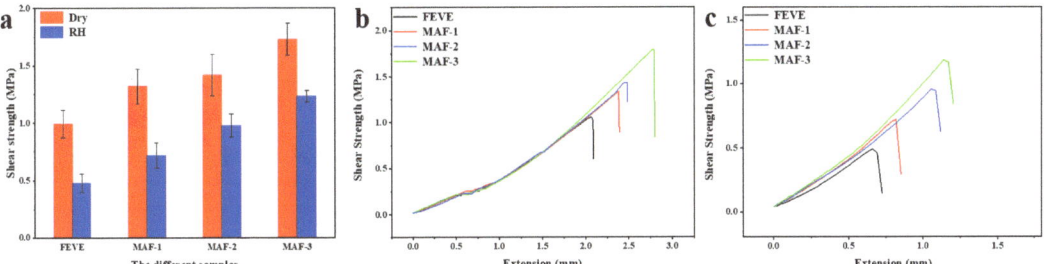

Figure 5. The shear strength of the wooden samples bonded with the MAF adhesives (**a**); stress–displacement curves for the wooden samples bonded using adhesives after drying (**b**) and after high RH degradation (**c**).

Table 1. Shear strength data for the wooden samples.

Adhesives	FEVE	MAF-1	MAF-2	MAF-3
Dry strength (MPa)	0.99	1.32	1.42	1.73
RH strength (MPa)	0.48	0.72	0.98	1.23

2.7. Antimicrobial Effects of Ag_2O/OH-MWCNTS-FEVE Adhesives

Figure 6 present the results of a comparative analysis of the antibacterial activity of the MAF-1, MAF-2, MAF-3, and FEVE adhesives, where all of the MAF adhesives exhibited inhibitory effects on three types of molds, i.e., *Trichoderma*, *Aspergillus niger*, and *white rot fungi*, where the best inhibitory effects were presented for Trichoderma (Figure 6). The images indicate that the inhibitory effects of MAF were dose-dependent, as the size of the inhibitory halo is directly proportional to the concentration of Ag_2O/OH-MWCNTS, with values exceeding 1 mm considered effective. According to established criteria, any antibacterial agent producing an inhibitory zone larger than 1 mm is classified as "good" [67]. Consequently, the incorporation of Ag_2O/OH-MWCNTS resulted in a notable enhancement of the antibacterial efficacy of FEVE adhesives.

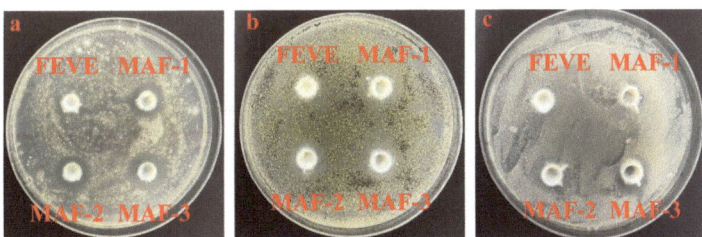

Figure 6. Inhibitory effects of different concentrations of MAF on (**a**) *Trichoderma*, (**b**) *Aspergillus niger*, and (**c**) *white rot fungi*.

3. Materials and Methods

3.1. Materials

$AgNO_3$ and NaOH were purchased from Sinopharm Chemical Reagent Co., Ltd. OH-MWCNTS was purchased from Xianfeng nanomaterial Technology Co., Ltd. (Jiangsu, China), Potato Dextrose Agar (PDA) was purchased from Aobox Biotechnology, *Trichoderma*, *Aspergillus niger*, and *white rot fungi* were purchased from Chinese Academy of Forestry (Beijing, China).

3.2. Preparation of Ag_2O/OH-MWCNTS-FEVE (MAF) Adhesives and Plywooden Samples

A total of 100 mg of OH-MWCNTs was dispersed in a 50 mL solution of 0.5 mol/L $AgNO_3$ (ethanol: deionized water = 1: 1) with sonication at a frequency of 10,000 Hz for 30 min. Subsequently, a 0.5 mol/L NaOH solution was incrementally added to the suspension of OH-MWCNTs/$AgNO_3$ in ice, with continuous stirring. Then, the resulting products were extracted, washed with deionized water and ethanol, and dried at 75 °C for 12 h, labelled as Ag_2O/OH-MWCNTs (MA), and stored in a desiccator until further use. 100 µg/mL, 200 µg/mL, and 300 µg/mL Ag_2O/OH-MWCNT-FEVE (MAF) were prepared and labelled as MAF1, MAF2, and MAF3, respectively. These solutions were subsequently sealed and stored for subsequent use.

The FEVE was diluted with ultrapure water to obtain a 50% aqueous solution and was labeled as FEVE.

3.3. Characterization

3.3.1. X-ray Diffraction Analysis

We weighed and characterized 10 mg Ag_2O/OH-MWCNTS and OH-MWCNTS powders using a high-resolution X-ray diffractometer (XRD, Smart Lab 9, Rigaku, Japan), with test conditions for Cu Kα ray (λ = 1.54056 A), a 2θ range of 20°–80°, acceleration voltage of 45 kV, tube current of 200 mA, and scanning speed of 5°/min.

3.3.2. Infrared Absorption Spectroscopy

We used 0.2 mg Ag_2O/OH-MWCNTS and OH-MWCNTS powders to prepare KBr pellets, and FTIR spectra were collected using the Fourier transform infrared (FTIR) spectrometer (Nicolet iS10, Thermo Fisher Scientific, Waltham, MA, USA). The spectral range was set between 4000 and 400 cm^{-1}, with a resolution of 4 cm^{-1}, and both the sample and background were scanned 64 times. The molecular structures of the samples were identified through the characteristic peaks of various functional groups in the spectra.

3.3.3. Scanning Electron Microscopy

The particles of Ag_2O/WCNTS were attached to the sample stage with the conductive adhesive, and the samples were sprayed with gold on the surface. Finally, the micro-morphology of the samples was obtained using a Field emission scanning electron microscope (SU8020, Hitachi Company, Japan).

3.3.4. Thermogravimetric Analysis

The thermal analysis of the three Ag_2O/OH-MWCNTS-FEVE adhesive samples (MAF-1, MAF-2, MAF-3) was performed with the Thermogravimetric analyzer (Themys One, Setaram, France) to evaluate its thermal stability. During the test, N_2 was used as the protective gas, the temperature range was set at 30–800 °C, and the heating rate was 20 °C/min.

3.3.5. Viscoelastic Measurement

The three Ag_2O/OH-MWCNTS-FEVE adhesive samples (MAF-1, MAF-2, MAF-3) were used and tested using the rheometer (MCR 302, Anton Paar, Austria). Rotors of PP50 were selected and 2 mL of the solution was taken with pipettes and placed on the test stage. The viscosity of the corresponding sample was measured in the shear rate range of 0.1–500 s^{-1}, and all measurements were made at a constant temperature of 23 °C.

3.3.6. Shear Strength

According to the standard ISO 6237:2003, adhesive should be applied to the surface of a wooden block sample with a coating area of 500 mm^2 (25 mm × 20 mm). Another wooden block of the same size should then be placed on top of the adhesive area, ensuring complete coverage, and clamped to secure. The assembled sample should be left at room temperature (23 ± 2 °C, relative humidity 50% ± 10) for 7 days. The assembled samples were used to measure the dry shear strength, or degradation under specific environmental conditions of 90% ± 5 relative humidity and 23 ± 2 °C for a duration of 5 days and tested with a high RH degradation shear strength test afterwards. The Universal Material Testing Machine (Kc-136pc, Xi'an Kaicheng Technology Co., Ltd., Xi'an, China) was utilized at a speed of 20 N/min for the test. This test was performed 10 times for each sample, and the average was calculated and used. The schematic diagram of the specimen for shear strength measurement is illustrated in Figure 7. Positioning of wooden samples on The Universal Material Testing Machine in Figure 8.

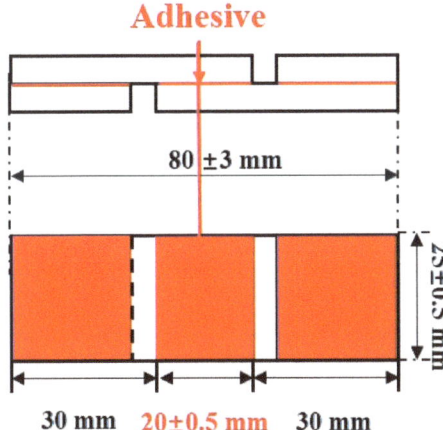

Figure 7. Schematic diagram of the test specimen used for shear strength measurement.

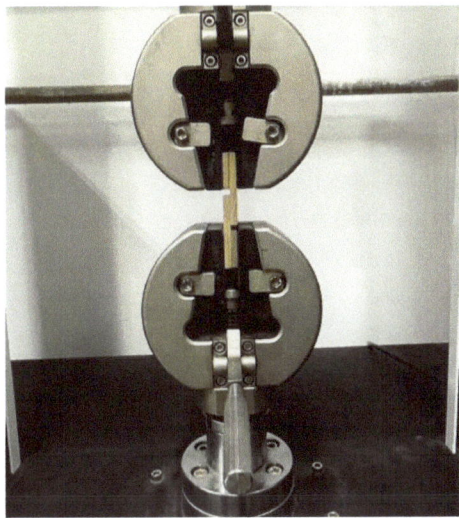

Figure 8. Positioning of a pair of wooden samples on The Universal Material Testing Machine.

3.3.7. Antifungal and Antibacterial Effects of Ag_2O/OH-MWCNTS-FEVE Adhesives

Aspergillus niger, *Trichoderma*, and *white rot fungi* were cultured in PDA inclined tube, then fungal spores were isolated with 0.05% Tween 20 and filtered with gauze. After counting mold spores with a blood cell counting plate, the spore suspension was diluted with phosphate-buffered saline (PBS) until 1×10^5 colony forming untis (CFU) was obtained. Then, 100 μL of 1×10^5 CFU/mL spore suspension was evenly applied to the PDA medium, and the medium was evenly divided into 4 parts along the center. An Oxford cup was placed in the center of each part, and 100 μL FEVE solution was added to the control group, and 100 μL MAF-1, MAF-2, MAF-3 solutions (100 μg/mL, 200 μg/mL, 300 μg/mL) were added to the treatment group successively, and cultured in an incubator at 25 °C with 80% RH for 48 h. The sizes of the fungal inhibition zones were observed and the images were collected.

4. Conclusions

Overall, a wood adhesive with elevated viscosity and effective bacteriostatic properties was synthesized through a straightforward method utilizing FEVE resin as the primary material given its superior viscosity and thermal resilience. Furthermore, the Ag_2O/OH-MWCNTS was prepared using an in situ precipitation technique, followed by the preparation of a composite adhesive by blending it with FEVE resin, denoted as MAF. The MAF adhesive displayed an exceptional thermal stability, with a lower mass loss rate than the original FEVE resin following treatment at 800 °C. At a consistent shear rate, the viscosity of MAF demonstrates a notable increase with the proportion of MA being higher than that of FEVE. This suggests that the nano-Ag_2O particles present in MA act as physical crosslinking agents in FEVE, enhancing the viscosity of the composite adhesive MAF. The adhesion strength between MAF and wood exhibits a similar trend, with wooden samples displaying higher shear strengths as the proportion of MA increases in comparison to FEVE. Additionally, the antibacterial effects of MAF adhesive were observed to be greater than 1mm for *Trichoderma*, *Aspergillus niger*, and *white rot fungi*. We also found that the antibacterial properties of the MAF adhesive were directly correlated with the concentration of Ag_2O/OH-MWCNTS, leading to the most effective inhibition of *Trichoderma*. Consequently, the MAF adhesive demonstrates significant promise for use as an adhesive for wooden artifacts, offering a novel approach for the rapid, environmentally friendly, and efficient preparation of composite adhesives with superior adhesion capabilities.

Author Contributions: Conceptualization and methodology, Y.L. (Yuhu Li), Y.L. (Yujia Luo), P.F. and G.T.; formal analysis and data curation, C.C., K.H., D.H. and J.L.; writing—original draft preparation, G.T.; writing—review and editing, Y.L. (Yujia Luo) and Y.L. (Yuhu Li). All authors have read and agreed to the published version of the manuscript.

Funding: The authors are grateful to Fundamental Research Funds for the Central Universities (GK202304013) and Shaanxi Key Research and Development Program of China (2024GX-YBXM-560).

Institutional Review Board Statement: Not applicable.

Informed Consent Statement: The study did not require ethical approval.

Data Availability Statement: The data presented in this study are included in the article.

Acknowledgments: The authors acknowledge the colleagues and collaborators who supported the research and the institutions and individuals who provided equipment support.

Conflicts of Interest: The authors declare no conflicts of interest.

References

1. Unger, A.; Schniewind, A.P.; Unger, W. *Conservation of Wooden Artifacts: A Handbook*; Springer Science & Business Media: Berlin/Heidelberg, Germany, 2001.
2. Peng, Y.; Wang, Y.; Zhang, R.; Wang, W.; Cao, J. Improvement of wooden against uv weathering and decay by using plant origin substances: Tannin acid and tung oil. *Ind. Crops Prod.* **2021**, *168*, 113606. [CrossRef]
3. Walsh-Korb, Z. Sustainability in heritage wooden conservation: Challenges and directions for future research. *Forests* **2022**, *13*, 18. [CrossRef]
4. Thieme, H. Lower palaeolithic hunting spears from germany. *Nature* **1997**, *385*, 807–810. [CrossRef]
5. Ambrose, S.H. Paleolithic technology and human evolution. *Science* **2001**, *291*, 1748–1753. [CrossRef]
6. Timar, M.C.; Sandu, I.C.A.; Beldean, E.C.; Sandu, I.G. Ftir investigation of paraloid b72 as consolidant for old wooden artefacts principle and methods. *Mater. Plast.* **2014**, *51*, 382–387.

7. Emmanuel, V.; Odile, B.; Céline, R. Ftir spectroscopy of woodens: A new approach to study the weathering of the carving face of a sculpture. *Spectrochim. Acta Part A Mol. Biomol. Spectrosc.* **2015**, *136*, 1255–1259. [CrossRef]
8. Lionetto, F.; Frigione, M. Mechanical and natural durability properties of wooden treated with a novel organic preservative/consolidant product. *Mater. Des.* **2009**, *30*, 3303–3307. [CrossRef]
9. Almkvist, G.; Persson, I. Fenton-induced degradation of polyethylene glycol and oak holocellulose. A model experiment in comparison to changes observed in conserved waterlogged wooden. *Holzforschung* **2008**, *62*, 704–708. [CrossRef]
10. Fors, Y.; Sandström, M. Sulfur and iron in shipwrecks cause conservation concerns. *Chem. Soc. Rev* **2006**, *35*, 399–415. [CrossRef] [PubMed]
11. Fors, Y.; Grudd, H.; Rindby, A.; Jalilehvand, F.; Sandström, M.; Cato, I.; Bornmalm, L. Sulfur and iron accumulation in three marine-archaeological shipwrecks in the baltic sea: The ghost, the crown and the sword. *Sci. Rep.* **2014**, *4*, 4222. [CrossRef]
12. Fors, Y.; Nilsson, T.; Risberg, E.D.; Sandström, M.; Torssander, P. Sulfur accumulation in pinewooden (pinus sylvestris) induced by bacteria in a simulated seabed environment: Implications for marine archaeological wooden and fossil fuels. *Int. Biodeterior. Biodegrad.* **2008**, *62*, 336–347. [CrossRef]
13. Fors, Y.; Jalilehvand, F.; Damian Risberg, E.; Björdal, C.; Phillips, E.; Sandström, M. Sulfur and iron analyses of marine archaeological wooden in shipwrecks from the baltic sea and scandinavian waters. *J. Archaeol. Sci. Rep.* **2012**, *39*, 2521–2532. [CrossRef]
14. Sandström, M.; Jalilehvand, F.; Damian, E.; Fors, Y.; Gelius, U.; Jones, M.; Salomé, M. Sulfur accumulation in the timbers of king henry viii's warship *mary rose*: A pathway in the sulfur cycle of conservation concern. *Proc. Natl. Acad. Sci. USA* **2005**, *102*, 14165–14170. [CrossRef]
15. Nagarajappa, G.B.; Pandey, K.K. Uv resistance and dimensional stability of wooden modified with isopropenyl acetate. *J. Photochem. Photobiol. B Biol.* **2016**, *155*, 20–27. [CrossRef] [PubMed]
16. Hon, D.N.S.; Chang, S.T. Surface degradation of wooden by ultraviolet light. *J. Polym. Sci. Polym. Chem. Ed.* **2003**, *22*, 2227–2241. [CrossRef]
17. Nair, S.; Nagarajappa, G.B.; Pandey, K.K. Uv stabilization of wooden by nano metal oxides dispersed in propylene glycol. *J. Photochem. Photobiol. B Biol.* **2018**, *183*, 1–10. [CrossRef] [PubMed]
18. Magdy, M. Analytical techniques for the preservation of cultural heritage: Frontiers in knowledge and application. *Crit. Rev. Anal. Chem.* **2022**, *52*, 1171–1196. [CrossRef] [PubMed]
19. Pike, R.A. Adhesive. Encyclopedia Britannica Online. 2022. Available online: https://www.britannica.com/technology/adhesive (accessed on 7 March 2024).
20. Chiavari, G.; Mazzeo, R. Characterisation of paint layers in chinese archaeological relics by pyrolysis-gc-ms. *Chromatographia* **1999**, *49*, 268–272. [CrossRef]
21. Calvano, C.D.; van der Werf, I.D.; Palmisano, F.; Sabbatini, L. Fingerprinting of egg and oil binders in painted artworks by matrix-assisted laser desorption ionization time-of-flight mass spectrometry analysis of lipid oxidation by-products. *Anal. Bioanal. Chem.* **2011**, *400*, 2229–2240. [CrossRef]
22. Luo, W.; Li, T.; Wang, C.; Huang, F. Discovery of beeswax as binding agent on a 6th-century bc chinese turquoise-inlaid bronze sword. *J. Archaeol. Sci.* **2012**, *39*, 1227–1237. [CrossRef]
23. Mitkidou, S.; Dimitrakoudi, E.; Urem-Kotsou, D.; Papadopoulou, D.; Kotsakis, K.; Stratis, J.A.; Stephanidou-Stephanatou, I. Organic residue analysis of neolithic pottery from north greece. *Microchim. Acta* **2008**, *160*, 493–498. [CrossRef]
24. Regert, M. Investigating the history of prehistoric glues by gas chromatography-mass spectrometry. *J. Sep. Sci.* **2004**, *27*, 244–254. [CrossRef]
25. Wei, S.; Pintus, V.; Pitthard, V.; Schreiner, M.; Song, G. Analytical characterization of lacquer objects excavated from a chu tomb in china. *J. Archaeol. Sci.* **2011**, *38*, 2667–2674. [CrossRef]
26. Unoki, M.; Kimura, I.; Yamauchi, M. Solvent-soluble fluoropolymers for coatings—Chemical structure and weatherability. *Surf. Coat. Int. Part B Coat. Trans.* **2002**, *85*, 209–213. [CrossRef]
27. Zhong, B.; Shen, L.; Zhang, X.; Li, C.; Bao, N. Reduced graphene oxide/silica nanocomposite-reinforced anticorrosive fluorocarbon coating. *J. Appl. Polym. Sci.* **2021**, *138*, 49689. [CrossRef]
28. Zhao, P.Y. Study on Preparation and Properties of Nano- Titanium Fluorocarbon Anticorrosion Coating. Master's Thesis, Harbin Institute of Technology, Harbin, China, 2010. (In Chinese with English Abstract)
29. Deacon, J. (Ed.) Fungal ecology: Saprotrophs. In *Fungal Biology*; Blackwell Publishing: Oxford, UK, 2005; pp. 213–236. [CrossRef]
30. Björdal, C.G.; Nilsson, T. Waterlogged archaeological wooden-a substrate for white rot fungi during drainage of wetlands. *Int. Biodeterior. Biodegrad.* **2002**, *50*, 17–23. [CrossRef]
31. Blanchette, R.A. A review of microbial deterioration found in archaeological wooden from different environments. *Int. Biodeterior. Biodegrad.* **2000**, *46*, 189–204. [CrossRef]
32. He, W.; Zhang, Y.; Li, J.; Gao, Y.; Luo, F.; Tan, H.; Wang, K.; Fu, Q. A novel surface structure consisting of contact-active antibacterial upper-layer and antifouling sub-layer derived from gemini quaternary ammonium salt polyurethanes. *Sci. Rep.* **2016**, *6*, 32140. [CrossRef] [PubMed]
33. Mohamed, G.G.; Soliman, M.H. Synthesis, spectroscopic and thermal characterization of sulpiride complexes of iron, manganese, copper, cobalt, nickel, and zinc salts. Antibacterial and antifungal activity. *Spectrochim. Acta Part A Mol. Biomol. Spectrosc.* **2010**, *76*, 341–347. [CrossRef] [PubMed]

34. Hameed, A.S.H.; Karthikeyan, C.; Ahamed, A.P.; Thajuddin, N.; Alharbi, N.S.; Alharbi, S.A.; Ravi, G. In vitro antibacterial activity of ZnO and nd doped ZnO nanoparticles against esbl producing escherichia coli and klebsiella pneumoniae. *Sci. Rep.* **2016**, *6*, 24312. [CrossRef]
35. Kasinathan, K.; Kennedy, J.; Elayaperumal, M.; Henini, M.; Malik, M. Photodegradation of organic pollutants rhb dye using uv simulated sunlight on ceria based TiO_2 nanomaterials for antibacterial applications. *Sci. Rep.* **2016**, *6*, 38064. [CrossRef] [PubMed]
36. Ma, S.; Zhan, S.; Jia, Y.; Zhou, Q. Superior antibacterial activity of Fe_3O_4-TiO_2 nanosheets under solar light. *ACS Appl. Mater. Interfaces* **2015**, *7*, 21875–21883. [CrossRef] [PubMed]
37. Zhu, Q.; Hu, X.; Stanislaus, M.S.; Zhang, N.; Xiao, R.; Liu, N.; Yang, Y. A novel $P/Ag/Ag_2O/Ag_3PO_4/TiO_2$ composite film for water purification and antibacterial application under solar light irradiation. *Sci. Total Environ.* **2017**, *577*, 236–244. [CrossRef] [PubMed]
38. Grandcolas, M.; Ye, J.; Hanagata, N. Combination of photocatalytic and antibacterial effects of silver oxide loaded on titania nanotubes. *Mater. Lett.* **2011**, *65*, 236–239. [CrossRef]
39. Sun, D.; Zhang, W.; Mou, Z.; Chen, Y.; Guo, F.; Yang, E.; Wang, W. Transcriptome analysis reveals silver nanoparticle-decorated quercetin antibacterial molecular mechanism. *ACS Appl. Mater. Interfaces* **2017**, *9*, 10047–10060. [CrossRef]
40. Rajabi, A.; Ghazali, M.J.; Mahmoudi, E.; Azizkhani, S.; Sulaiman, N.H.; Mohammad, A.W.; Mustafah, N.M.; Ohnmar, H.; Naicker, A.S. Development and antibacterial application of nanocomposites: Effects of molar ratio on Ag_2O–CuO nanocomposite synthesised via the microwave-assisted route. *Ceram. Int.* **2018**, *44*, 21591–21598. [CrossRef]
41. Tripathi, S.; Mehrotra, G.K.; Dutta, P.K. Chitosan–silver oxide nanocomposite film: Preparation and antimicrobial activity. *Bull. Mater. Sci.* **2011**, *34*, 29–35. [CrossRef]
42. Hu, Z.; Chan, W.L.; Szeto, Y.S. Nanocomposite of chitosan and silver oxide and its antibacterial property. *J. Appl. Polym. Sci* **2008**, *108*, 52–56. [CrossRef]
43. Trang, V.T.; Tam, L.T.; Van Quy, N.; Huy, T.Q.; Thuy, N.T.; Tri, D.Q.; Cuong, N.D.; Tuan, P.A.; Van Tuan, H.; Le, A.-T.; et al. Functional iron oxide–silver hetero-nanocomposites: Controlled synthesis and antibacterial activity. *J. Electron. Mater.* **2017**, *46*, 3381–3389. [CrossRef]
44. Sen, R.; Govindaraj, A.; Rao, C.N.R. Carbon nanotubes by the metallocene route. *Chem. Phys. Lett.* **1997**, *267*, 276–280. [CrossRef]
45. Moreno, V.; Llorent-Martínez, E.J.; Zougagh, M.; Ríos, A. Decoration of multi-walled carbon nanotubes with metal nanoparticles in supercritical carbon dioxide medium as a novel approach for the modification of screen-printed electrodes. *Talanta* **2016**, *161*, 775–779. [CrossRef]
46. Shi, L.; Zhang, G.; Wang, Y. Tailoring catalytic performance of carbon nanotubes confined CuO-CeO_2 catalysts for CO preferential oxidation. *Int. J. Hydrogen Energy* **2018**, *43*, 18211–18219. [CrossRef]
47. Obaidullah, I. Carbon nanotube membranes for water purification: Developments, challenges, and prospects for the future. *Sep. Purif. Technol.* **2019**, *209*, 307–337. [CrossRef]
48. Jha, R.; Singh, A.; Sharma, P.K.; Fuloria, N.K. Smart carbon nanotubes for drug delivery system: A comprehensive study. *J. Drug Deliv. Sci. Technol.* **2020**, *58*, 101811. [CrossRef]
49. Wang, P.; Gao, S.; Chen, X.; Yang, L.; Wu, X.; Feng, S.; Hu, X.; Liu, J.; Xu, P.; Ding, Y. Effect of hydroxyl and carboxyl-functionalized carbon nanotubes on phase morphology, mechanical and dielectric properties of poly(lactide)/poly(butylene adipate-co-terephthalate) composites. *Int. J. Biol. Macromol.* **2022**, *206*, 661–669. [CrossRef] [PubMed]
50. Shimizu, T.; Ding, W.; Kameta, N. Soft-matter nanotubes: A platform for diverse functions and applications. *Chem. Rev.* **2020**, *120*, 2347–2407. [CrossRef] [PubMed]
51. Kang, S.; Mauter, M.S.; Elimelech, M. Physicochemical determinants of multiwalled carbon nanotube bacterial cytotoxicity. *Environ. Sci. Technol.* **2008**, *42*, 7528–7534. [CrossRef] [PubMed]
52. Zhang, Z.; Qi, Y.; Zhai, H. *Surface Characteristics and Electric Conductivity of Mwcnts/Feve Copolymer Composite Coatings*; Springer International Publishing: Cham, Switzerland, 2016; pp. 2117–2122.
53. Epp, J. 4—X-ray diffraction (xrd) techniques for materials characterization. In *Materials Characterization Using Nondestructive Evaluation (Nde) Methods*; Hübschen, G., Altpeter, I., Tschuncky, R., Herrmann, H.-G., Eds.; Woodenhead Publishing: Cambridge, UK, 2016; pp. 81–124.
54. Ananth, A.; Mok, Y.S. Dielectric barrier discharge (DBD) plasma assisted synthesis of Ag_2O nanomaterials and Ag_2O/RuO_2 nanocomposites. *Nanomaterials* **2016**, *6*, 42. [CrossRef]
55. Sarode, V.B.; Patil, R.D.; Chaudhari, G.E. Characterization of functionalized multi-walled carbon nanotubes. *Mater. Today Proc.* **2023**. [CrossRef]
56. Ravichandran, S.; Paluri, V.; Kumar, G.; Loganathan, K.; Kokati Venkata, B.R. A novel approach for the biosynthesis of silver oxide nanoparticles using aqueous leaf extract of *callistemon lanceolatus* (myrtaceae) and their therapeutic potential. *J. Exp. Nanosci.* **2016**, *11*, 445–458. [CrossRef]
57. Lucacel, R.C.; Marcus, C.; Timar, V.; Ardelean, I.I. Ft-ir and raman spectroscopic studies on B_2O_3-PbO-Ag_2O glasses dopped with manganese ions. *Solid State Sci.* **2007**, *9*, 850–854. [CrossRef]
58. Hootifard, G.; Sheikhhosseini, E.; Ahmadi, S.A.; Yahyazadehfar, M. Synthesis and characterization of CO-MOF@Ag_2O nanocomposite and its application as a nano-organic catalyst for one-pot synthesis of pyrazolopyranopyrimidines. *Sci. Rep.* **2023**, *13*, 17500. [CrossRef] [PubMed]

59. Li, D.; Chen, S.; Zhang, K.; Gao, N.; Zhang, M.; Albasher, G.; Shi, J.; Wang, C. The interaction of Ag_2O nanoparticles with *Escherichia coli*: Inhibition–sterilization process. *Sci. Rep.* **2021**, *11*, 1703. [CrossRef]
60. Çakır, Ü.; Kestel, F.; Kızılduman, B.K.; Bicil, Z.; Doğan, M. Multi walled carbon nanotubes functionalized by hydroxyl and schiff base and their hydrogen storage properties. *Diam. Relat. Mater.* **2021**, *120*, 108604. [CrossRef]
61. Liu, H.; Li, C.; Sun, X.S. Soy-oil-based waterborne polyurethane improved wet strength of soy protein adhesives on wooden. *Int. J. Adhes. Adhes.* **2017**, *73*, 66–74. [CrossRef]
62. Fang, Y.; Qin, L.; Zhao, W.; Qin, B.; Zhang, X.; Wu, X. Research progress and development trend on corrosion resistant fluorocarbon paint. *J. Chin. Soc. Corros. Prot.* **2016**, *36*, 97–106.
63. Shi, Y.; Wang, G. Influence of molecular weight of peg on thermal and fire protection properties of pepa-containing polyether flame retardants with high water solubility. *Prog. Org. Coat.* **2016**, *90*, 390–398. [CrossRef]
64. Mo, M.; Zhao, W.; Chen, Z.; Yu, Q.; Zeng, Z.; Wu, X.; Xue, Q. Excellent tribological and anti-corrosion performance of polyurethane composite coatings reinforced with functionalized graphene and graphene oxide nanosheets. *RSC Adv.* **2015**, *5*, 56486–56497. [CrossRef]
65. Li, M.; Liu, Q.; Jia, Z.; Xu, X.; Cheng, Y.; Zheng, Y.; Xi, T.; Wei, S. Graphene oxide/hydroxyapatite composite coatings fabricated by electrophoretic nanotechnology for biological applications. *Carbon* **2014**, *67*, 185–197. [CrossRef]
66. Mišković-Stanković, V.; Jevremović, I.; Jung, I.; Rhee, K. Electrochemical study of corrosion behavior of graphene coatings on copper and aluminum in a chloride solution. *Carbon* **2014**, *75*, 335–344. [CrossRef]
67. Alahmadi, N.S.; Betts, J.W.; Cheng, F.; Francesconi, M.G.; Kelly, S.M.; Kornherr, A.; Prior, T.J.; Wadhawan, J.D. Synthesis and antibacterial effects of cobalt–cellulose magnetic nanocomposites. *RSC Adv.* **2017**, *7*, 20020–20026. [CrossRef]

Disclaimer/Publisher's Note: The statements, opinions and data contained in all publications are solely those of the individual author(s) and contributor(s) and not of MDPI and/or the editor(s). MDPI and/or the editor(s) disclaim responsibility for any injury to people or property resulting from any ideas, methods, instructions or products referred to in the content.

Article

Ionic Liquids as Reconditioning Agents for Paper Artifacts

Catalin Croitoru * and Ionut Claudiu Roata

Materials Engineering and Welding Department, Transilvania University of Brasov, Eroilor 29 Str., 500039 Brasov, Romania; ionut.roata@unitbv.ro
* Correspondence: c.croitoru@unitbv.ro

Abstract: This research explores the potential of ionic liquids (ILs) in restoring paper artifacts, particularly an aged book sample. Three distinct ILs—1-ethyl-3-propylimidazolium bis(trifluoromethylsulfonyl)imide, 1-methyl-3-pentylimidazolium bis(trifluoromethylsulfonyl)imide, and 1-methyl-3-heptylimidazolium bis(trifluoromethylsulfonyl)imide —both in their pure form and isopropanol mixtures, were examined for their specific consumption in conjunction with paper, with 1-ethyl-3-propylimidazolium bis(trifluoromethylsulfonyl)imide displaying the highest absorption. Notably, the methyl-3-heptylimidazolium ionic liquid displayed pronounced deacidification capabilities, elevating the paper pH close to a neutral 7. The treated paper exhibited significant color enhancements, particularly with 1-heptyl-3-methylimidazolium and 1-pentyl-3-methylimidazolium ILs, as evidenced by CIE-Lab* parameters. An exploration of ILs as potential UV stabilizers for paper unveiled promising outcomes, with 1-heptyl-3-methylimidazolium IL demonstrating minimal yellowing post-UV irradiation. FTIR spectra elucidated structural alterations, underscoring the efficacy of ILs in removing small-molecular additives and macromolecules. The study also addressed the preservation of inked artifacts during cleaning, showcasing ILs' ability to solubilize iron gall ink, particularly the one with the 1-ethyl-3-propylimidazolium cation. While exercising caution for prolonged use on inked supports is still recommended, ILs are shown here to be valuable for cleaning ink-stained surfaces, establishing their effectiveness in paper restoration and cultural heritage preservation.

Keywords: paper artifacts; cellulose; ionic liquids; cleaning agents; preservation

Citation: Croitoru, C.; Roata, I.C. Ionic Liquids as Reconditioning Agents for Paper Artifacts. *Molecules* **2024**, *29*, 963. https://doi.org/10.3390/molecules29050963

Academic Editors: Yueer Yan, Yi Tang and Yuliang Yang

Received: 27 January 2024
Revised: 14 February 2024
Accepted: 20 February 2024
Published: 22 February 2024

Copyright: © 2024 by the authors. Licensee MDPI, Basel, Switzerland. This article is an open access article distributed under the terms and conditions of the Creative Commons Attribution (CC BY) license (https://creativecommons.org/licenses/by/4.0/).

1. Introduction

Preserving and restoring cellulose paper artifacts, such as books, is of utmost importance in order to protect our cultural heritage and ensure their longevity [1–3]. However, over time, cellulose artifacts are subject to various processes that can lead to degradation [4,5]. Understanding degradation and implementing effective restoration methods is crucial for the preservation of these materials [6]. The altering of cellulose paper involves intricate mechanisms, often stemming from a combination of factors such as exposure to microorganisms, light, humidity, temperature fluctuations, and pollutants [7–9]. Endogenous factors, such as the inherent quality of cellulose fibers, the presence of lignin, metallic ions, and various compounds used as sizers and binders, further contribute to the degradation process [5,10,11]. These factors collectively lead to physical damage, discoloration, staining, and acidification, ultimately compromising the structural integrity of the paper [10].

Traditional methods for restoration involve color restoration (including the removal of various types of stains), and deacidification. Color restoration aims to revive the original vibrancy of the paper, while deacidification helps neutralize the acidity that can accelerate degradation [12–14].

Color restoration and cleaning techniques can be time-consuming multi-step processes and may involve the use of chemicals such as hydrogen peroxide, chlorine or hypochlorites [15], enzymes [16] or even gamma radiation [17] and laser beams [18] that can be harsh on the paper leading to alterations in its mechanical properties and original appearance. Additionally, attempts to use mechanical cleaning methods, such as erasing and abrasive cleaning, may result in damage to the paper [19]. Post-production deacidification methods, such as non-aqueous solvent cleaning or spraying with alkaline preserves, may not always effectively remove all acidity or can cause structural damage to cellulose in the long term [14,20,21].

Irrespective of the chosen restoration methods and techniques, they must adhere to fundamental criteria, with paramount importance given to increased efficiency and minimal potential for paper substrate degradation [3,22]. Preserving the original appearance and texture of the material, ensuring long-term stability for any introduced additives or consolidants, and maintaining environmental friendliness with low toxicity are equally pivotal considerations [23]. These criteria collectively form the essential framework for evaluating the efficacy and sustainability of any restoration approach, emphasizing the need for holistic and conscientious practices in the preservation of cultural heritage materials. A critical analysis of the restoration and conservation methods to date for cellulosic materials shows that all traditional methods basically cannot meet all these criteria simultaneously [23,24]. Although these traditional methods have proven effective in a strict sense, such as cost efficiency or end result, they certainly have drawbacks with regard to other aspects.

These challenges highlight the need for alternative restoration methods that can address these limitations and provide a more effective, gentle, and sustainable approach to preserving cultural heritage [25].

One potential solution that has emerged in recent years is the use of ionic liquids (ILs) in conjunction with artifact preservation [25–27]. Ionic liquids are composed entirely of ions, which gives them unique properties that make them suitable for restoration purposes. Their cleaning and color restoration effect is based on the ability to partly dissolve cellulose and other macro- or small-molecular compounds allowing for the removal of stains and contaminants [28–30]. They can also be used to impart antibacterial or antifungal character to the paper surface for improved preservation by disrupting the cell wall membranes or by interfering with the metabolism of degrading microorganisms [23,31]. In addition to these properties, they also present low vapor pressure, high chemical stability, tunability and recyclability, which allows for the optimization and reuse of IL-based reconditioning formulations. However, ILs can also dissolve and damage cellulose, the main component of paper, which can result in the loss of mechanical and structural integrity of paper [31,32]. Therefore, there is a tradeoff between the cleaning and solvency effects of ILs, which depends on the type and concentration of ILs, the duration and temperature of the treatment, and the composition and condition of the paper. To minimize the solvency effect of ILs on cellulose and maximize the cleaning effect of ILs on paper, it is preferable to use ILs with longer alkyl chain cations or ILs with anions that do not have a strong impact on cellulose [32].

There are limited studies in the literature on the direct application of ILs on old and degraded book cellulose substrates. Most of the information available is indirect, and is on their application as wood preservatives, in cellulose dissolution and recovery, as topical biocides, or as solvents for various systems [23,26,33]. Although their long-term impact on artifacts is not yet completely understood, the specific use of ionic liquids in cellulose preservation has shown promising results. Ionic liquids have the ability to stabilize the pH of the paper, preventing further degradation [23,30]. Ionic liquids with short side alkyl chains such as 1-butyl-3-methylimidazolium benzotriazole and 1-butyl-3-methylimidazolium 1,2,4-triazolate showed the highest deacidification capability, increasing the paper pH to more than 7 with solutions at concentrations above 3%, but caused modifications in paper texture and transparency [23].

Also, due to their plasticizing effect on cellulose and lignin, ILs can improve the mechanical resistance of paper artifacts by increasing their tensile strength, fracture toughness, elongation, burst index, tear index, and fold endurance [34–36].

Ionic liquids, particularly protic ionic liquids (PILs) and those with azole-functionalities, have been found to exhibit significant biocidal and antibacterial properties when used in the restoration of cellulose paper in artifacts. PILs with 1-ammonium-2-propanol cation and various anions have shown antimicrobial activity against a range of microorganisms, including fungi, bacteria, and yeasts [37]. Furthermore, the addition of specific ionic liquids, such as benzalkonium nitrate(V), benzalkonium DL-lactate, and didecylodimethylammonium DL-lactate, to paper samples or pine bleached kraft pulp has been shown effective against paper-infesting molds and yeasts [23,27,31].

In this study, we explore novel approaches to preserving and restoring cellulose paper artifacts, with a specific focus on books, aiming to overcome the challenges posed by their inevitable degradation over time. The study utilizes three ionic liquids, specifically 1-ethyl-3-propylimidazolium bis(trifluoromethylsulfonyl)imide (PrEIMTs), 1-methyl-3-pentylimidazolium bis(trifluoromethylsulfonyl)imide (PentMIMTs), and 1-heptyl-3-methylimidazolium bis(trifluoromethylsulfonyl)imide (HeptMIMTs). By subjecting paper strips from an aged book to short-term immersion treatments in IL/isopropyl alcohol mixtures, the study systematically assesses the impact of these treatments on the paper color restoration, stain removal and deacidification. The chosen ILs are able to clean the paper supports and practically completely deacidify them in much shorter time periods (6 h) than those reported in the literature (typically over 24 h), without modification in the paper's texture or opacity. Additionally, the research investigates the potential of ILs to act as UV stabilizers, offering protection against UV-induced color alterations, a novel application not extensively explored in previous literature. Moreover, the study delves into the interaction between ILs and iron gall ink, a common ink associated with paper artifacts [38], demonstrating the compatibility of IL treatment with inked paper supports, which is crucial for preserving handwritten or printed content during restoration processes. Overall, by addressing these aspects, this study contributes to the advancement of restoration techniques for cellulose paper artifacts, offering a more effective, gentle, and sustainable approach to preserving cultural heritage materials.

2. Results and Discussion

The specific consumption (S_c) values of ionic liquids from both mixtures with isopropanol (10% v/v and 50% v/v), as well as those of pure ILs, for Whatman filter paper, are depicted in Figure 1. Typically, S_c increases with the IL amount and is also influenced by the molar mass of the IL. Consequently, the PrEIMTs ionic liquid is absorbed in the highest amount.

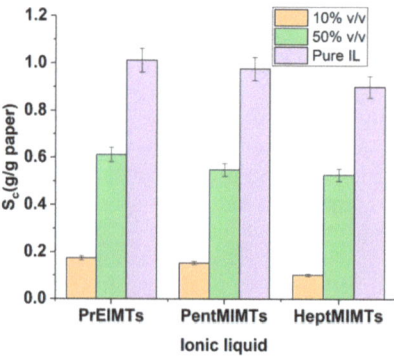

Figure 1. The specific consumption (S_c) of pure ILs and ILs/isopropanol mixtures for Whatman filter paper.

This absorption trend remains consistent, and the same pattern is evident for the B1 paper sample, particularly in the case of the 10% v/v mixture with isopropanol, as shown in Table 1. It is worth noting that the values are slightly lower, attributable to the higher compactness of the book paper. Information on the specific consumption of ILs for paper treatment in the reference literature is limited and inconsistently reported. For the sake of comparison, it is noted that specific consumption values for 1-butyl-3-methylimidazolium ILs with triazole anions were within the range of 0.5 to 2 g/g of paper, but these values were obtained with a 24 h treatment duration [23].

Table 1. Specific consumption (S_c) of paper restauration agents used in this study at the concentration of 10% v/v and the values supporting the deacidification of paper and changes in ash content.

Sample	S_c (g/g Paper)	Paper pH	Ash Content (%)	Paper Basis Weight (g/m^2)
B1-PrEIMTs	0.17 ± 0.03	6.50 ± 0.20	0.93 ± 0.15	64 ± 0.25
B1-PentMIMTs	0.14 ± 0.01	6.70 ± 0.15	1.02 ± 0.18	65 ± 0.15
B1-HeptMIMTs	0.09 ± 0.01	7.00 ± 0.10	1.01 ± 0.15	65 ± 0.30

The treatment of the B1 paper revealed that the ionic liquid HeptMIMTs exhibited a good capability for deacidification of paper. The 10% solution of this IL in isopropanol markedly increased the pH to 7.00, compared to the initial pH of 4.30. The other ILs caused a marginally lower shift in pH, but they could be deemed efficient for this purpose as well. Similar effective deacidification of old paper has been reported for small-chain alkylimidazolium halides but for a longer treatment duration [23].

There are several mechanisms that can cause paper acidification, such as the partial degradation or hydrolysis of acidic raw materials used during the papermaking process, which degrade over time and form organic acids, such as acetic, formic, or oxalic acids, or inorganic acids, such as sulfuric or aluminum sulfate acids, that can lower the pH of paper. Another mechanism involves acid hydrolysis that occurs when paper is exposed to moisture and heat, which breaks the bonds between the glucose units of cellulose, the main component of paper, and releases protons. The protons then react with the remaining cellulose chains, forming more water and reducing the pH of paper.

Ionic liquids can deacidify paper by dissolving and extracting various organic and inorganic compounds from paper, such as lignin, hemicellulose, cellulose, metal ions, and salts, which can reduce the amount and therefore effect of acidic species in the paper and increase its pH [39]. This is the main mechanism of paper deacidification by Ils, as evidenced by the FTIR studies (for organic species) and by the lower ash content of the supports treated with Ils compared to B1 (for inorganic species) (Table 1). Also, there seems to be a slight reduction in the paper's basis weight of the treated supports compared to B1, which could further sustain the washing effect of the ILs on paper. However, other mechanisms may also contribute to the deacidification effect, such as the hydrogen bonding between ILs and cellulose, which may alter the acidity and basicity of the hydroxyl groups, and the other types of interactions between ILs and paper, such as electrostatic, van der Waals, or π–π stacking interactions, which may influence the solubility and stability of paper components [36,40]. The ILs used in this study have the same anion, and similar cations, which are N-alkyl-N'-methylimidazolium derivatives with different alkyl chain lengths. The sulfonylimide anion is a weak base, with a pKa of about 5.5, and it is unlikely to accept a proton from a weak acid, such as the carboxylic acids that could be present in the paper [41]. Moreover, the imide anion is highly polarizable and delocalized, which reduces its basicity and increases its interaction with the ILs cation. Therefore, the reaction between the imide anion and the protons in the paper is probably negligible or insignificant. However, the cation and the alkyl chain length may play a role in the interaction with paper, as they affect the solvation, dispersion, and penetration of the ILs into the paper structure.

Figure 2 illustrates the modifications brought to the paper by the IL-cleaning process studied here, while Figure 3a–c provide the quantitative CIE-Lab* parameters to assess those modifications and their interpretation.

Figure 2. Photographic images of the initial B1 paper strips, of the same strips cleaned with IL/isopropanol solutions and IL-treated strips submitted to accelerated UVA aging (254 nm).

The B1 paper material exhibits a noticeable yellow/reddish hue, as evidenced by positive values in the a* and b* parameters (Figure 3a,b). The immersion of the paper in 10% IL solutions in isopropanol effectively "washes" the paper, restoring whiteness (by up to 6%) and eliminating a portion of the respective tinges that contribute to the paper's aged/degraded appearance (a decrease of up to 20% in a* values and 55% in b* values, as shown in Figure 3a). Notably, the highest color differences are observed for HeptMIMTs and PentMIMTs ILs, in Figure 3c.

To assess as much as possible all the benefits of the IL treatment of aged paper, the possibility of extending the applicability of these compounds as UV stabilizers for this type of material was studied. The effect of color changes on IL-treated paper samples and on B1 reference under UVA irradiation aging is shown in Figures 2 and 3a–c. While irradiation could drastically affect the color of the untreated material, the modifications in color of the IL-treated samples are less pronounced. With irradiation, the color of the B1 paper samples became significantly darker (a decrease in the lightness parameter, L*) and bluish/greenish, consistent with other findings [42,43].

The UV-aged IL-treated paper strips actually presented a minimal increase in lightness (under 5%), probably due to the ILs inducing glossiness to the surface [43] and restructuring the material, coupled with a yellowing/reddening of the material. For the sample treated with HeptMIMTs, only a 1% in yellowing and 2% total color modifications were observed from the b* values and ΔE*, respectively, possibly indicating that this IL would be best suited for the UV stabilizing of paper. This is consistent with our previous studies on wood and cellulose fibers, where the IL with the highest lateral alkyl chain also proved to be the most effective in inhibiting the formation of free radicals and chromophore groups responsible for yellowing/reddening [34,44,45].

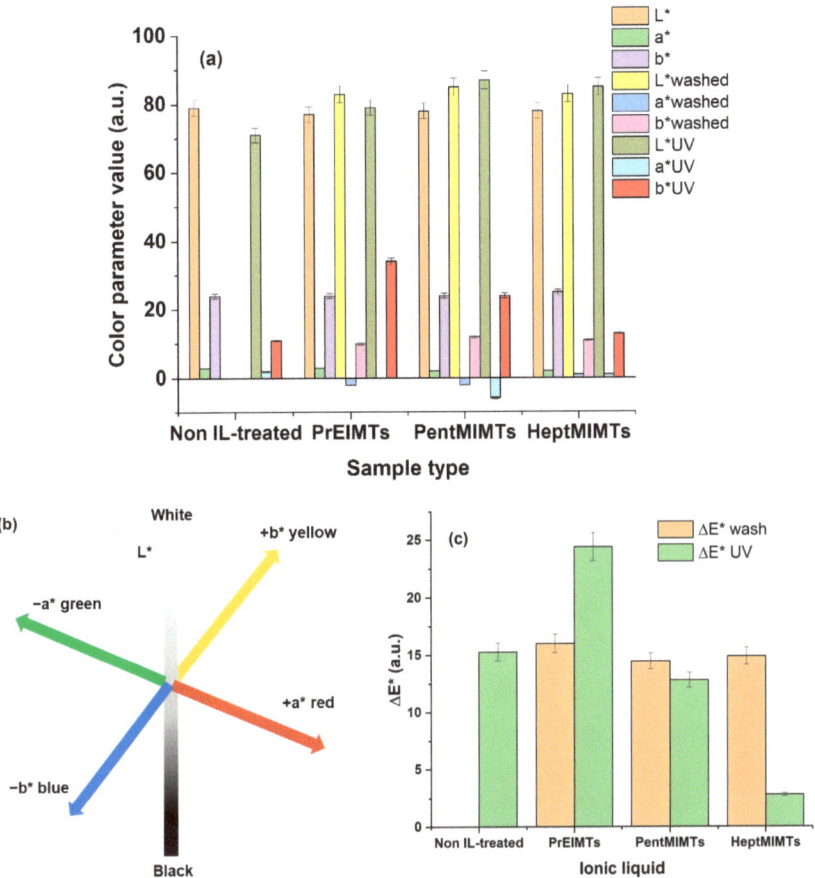

Figure 3. (**a**) CIE-Lab* color parameters for the IL-treated paper samples and for the samples submitted to UV aging; (**b**) CIE-Lab* space color coordinates; (**c**) total color modifications for the washing step and accelerated UV aging.

As evidenced from the optical microscopy images (Figure 4), the fibers in the untreated paper (B1) appear compact and closely knit (probably due to the binder), indicating that they have a small diameter and a high density. The fibers in the paper treated with PrEIMTs (B1-PrEIMTs) appear slightly more separated and cleaner than B1, suggesting that the ionic liquid has dissolved and removed some of the stains, dirt, and degraded compounds from the paper surface. Also, this ionic liquid has caused swelling of the fibers, increased their diameter and reduced their density to the highest extent.

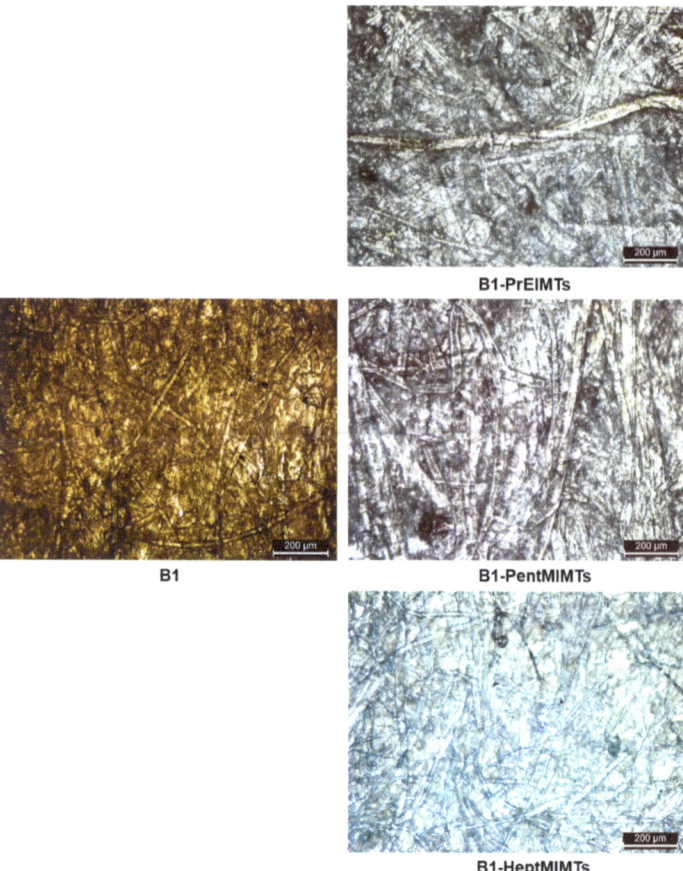

Figure 4. Optical microscopy images of paper strips used in the washing treatment of B1 paper substrates with IL 10% v/v solutions (100×).

The fibers in the paper treated with PentMIMTs (B1-PentMIMTs) show more distinct and visible fibers than B1-PrEIMTs, indicating that the ionic liquid has a stronger cleaning effect than PrEIMTs, but with some variation or inconsistency.

The fibers in the paper treated with HeptMIMTs (B1-HeptMIMTs) also show distinct fibers, but with less clarity than B1-PentMIMTs. This may suggest that the ionic liquid has a similar cleaning or swelling effect as PentMIMTs. The fibers have a comparable diameter and density as B1-PentMIMTs, and the red-brownish grime was mobilized to the highest extent.

This variation in the cleaning effect of the ILs suggest that their efficiency in this case is not only influenced by their solvency effect, but other factors may play a role as well, such as their surface tension, as higher molar mass ILs are known to possess a surfactant character [46].

The swelling effect is the expansion of the cellulose fibers due to the penetration of the IL molecules into the fiber structure. The average cellulose fiber diameters increased after the IL treatment, in the order: B1 (12.5 μm) < B1-HeptMIMTs (14.3 μm) < B1-PentMIMTs (15.4 μm) < B1-PrEIMTs (17.5 μm). This order corresponds to the decreasing crystallinity index and the increasing hydrogen bond disruption ability of the ILs, as confirmed by the FTIR spectra analysis. The increase in diameter due to fiber swelling ranged from 13% for HeptMIMTs to 18% for PentIMIMTs. This increase is typical also for the paper industry (pulp refining increases the initial fiber diameters by 5–10%) [47], as well as for the domain of paper restoration. For example, the fiber diameter can increase by 10–20% after alkaline treatment, depending on the concentration and duration of the treatment [48].

The distinctive features of the IL washing agents in their FTIR spectra from Figure 5a manifest primarily in two regions: the CH stretching region (2800–3180 cm^{-1}) and the bond-stretching/fingerprint region (below 1800 cm^{-1}). Most notably, in the -CH stretching region, the band at 3150–3170 cm^{-1} corresponds to the symmetric stretching of H-C-C-H in the imidazolium ring, along with stretching of the N(CH)N C-H ring [49,50]. Within the bond-stretching/fingerprint region, notable bands for the cation include those at ~1565 cm^{-1} (C=C stretch), at ~1132 cm^{-1} (-N-CH$_3$ twisting) and several weak overtones below 800 cm^{-1} [51]. The Ts anion contributes strong bands at ~785 cm^{-1} and 1185 cm^{-1}, associated with CF$_3$ symmetric bending and stretching. Other bands at 1045 cm^{-1} and 1335 cm^{-1} indicate SO$_2$ stretching, complemented by several overtones of medium intensity below 800 cm^{-1} [52]. The lowest molar mass IL, PrEIMTs, exhibits traces of adsorbed water, evidenced by -OH stretching vibrations at ~3360 cm^{-1}.

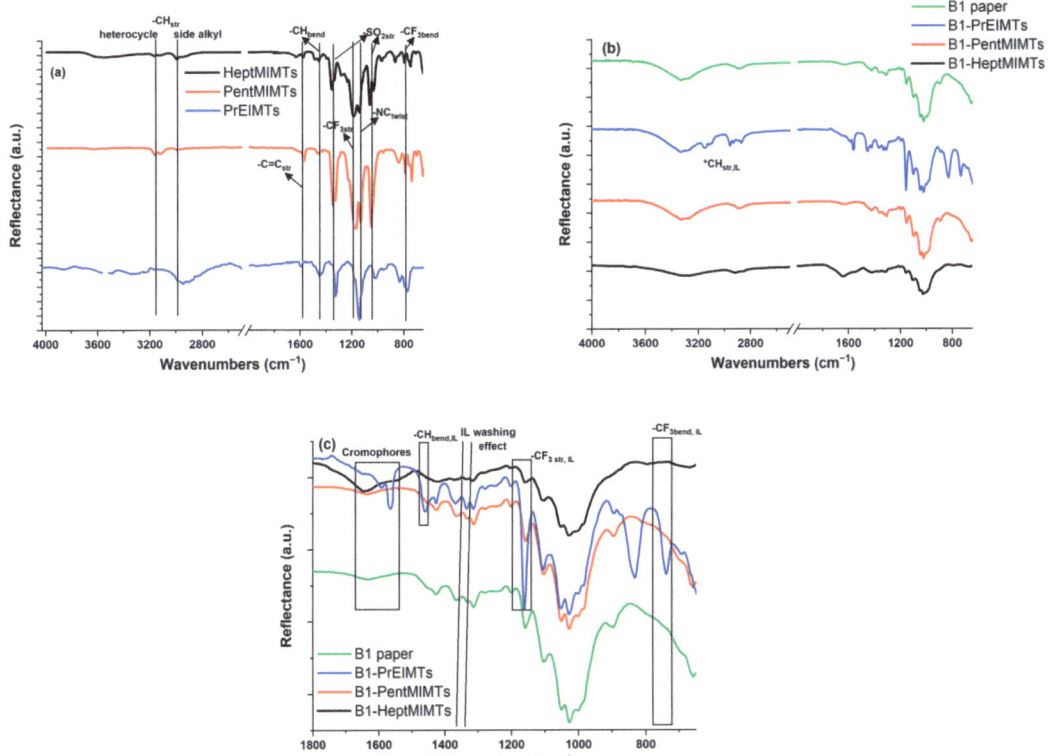

Figure 5. FTIR spectra of (**a**) ILs; (**b**) B1 paper strip and B1 strips treated with ILs; (**c**) zoom in the fingerprint region of B1 paper strip and B1 strips treated with ILs.

Following treatment with the three ionic liquids (ILs), the respective FTIR spectra of B1-PrEIMTs, B1-PentMIMTs, and B1-HeptMIMTs samples exhibit distinctive bands associated with both the IL cation (Figure 5b, imidazolium -CH stretch) and, more prominently, the anion (Figure 4c, -CF$_3$ stretch modes).

To evaluate the structural changes in the paper supports, the crystallinity index for cellulose was estimated by calculating the hydrogen bond index (HBI), which is a ratio of the intensity of the O-H stretching band at 3315 cm^{-1} to the intensity of the -CH and -CH$_2$

bending band at 1315 cm^{-1}. The HBI reflects the degree of hydrogen bonding between the hydroxyl groups of cellulose, which is lower in the crystalline region and higher in the amorphous region. Therefore, the higher the HBI, the lower the crystallinity index, as more hydroxyl groups are involved in hydrogen bonding in the amorphous region, which results from the transformation of cellulose I (crystalline) to cellulose II (amorphous) [53].

The IL treatment of the B1 sample at the relatively low concentration of 10% v/v in this study appears to have a moderate impact on the material's structure. No additional bands, beyond those associated with the main components in paper and the IL, were identified. The overall shape of the spectra and the relative ratio between the main bands attributed to the cellulose structural backbone, such as the -C-O-C- α-1,4 glycosidic linkages (1149 cm^{-1}) and the glycosidic ring (1078 cm^{-1}) [43], appear largely unchanged.

The HBI values decrease in the order B1-PrMIMTs (4.68) > B1-PentMIMTs (4.20) > B1 (4.00) > B1-HeptMIMTs (3.95), indicating that the lower mass IL (PrMIMTs) had the highest cellulose hydrogen bond disruption potential.

There are clear indications that the treatment promotes the removal of small-molecular (partly degraded) paper additives, macromolecules (cellulose, lignin) and acidic species due to the effective solvation ability of the ILs. For example, in the chromophore group region highlighted in Figure 5c, the aged paper exhibits weak stretching vibration modes specific to conjugated C=O in xylans (1582 cm^{-1}) and C=C stretching in lignin (1600 cm^{-1}) [54,55]. Conversely, an increase in the intensity of bands associated with lignin and xylan was observed for B1-PrEIMTs, B1-PentMIMTs, and B1-HeptMIMTs, likely due to the ILs' structure-mobilizing effect. The ILs seem to facilitate the removal of partly degraded macromolecules contributing to paper yellowing, while the restructuring of cellulose exposes a higher number of bonds belonging to both lignin and xylan units to the IR beam, maintaining or even enhancing the overall band intensity.

The FTIR spectra in Figure 5c also provide insights into the nature of paper additives. The broad absorption at ~1635 cm^{-1} in B1 could be attributed to amide I bands, potentially from gelatin [56,57], while weak bands at 1357 and 1329 cm^{-1} may be associated with fatty acid salts like stearates (commonly used as paper additives) or even lipids characteristic of molds or fungi [58]. All three ILs appear to mobilize these small-molecular compounds, known to accumulate dirt and contribute to yellowing over time, leading to the partial restoration of paper color, as observed in Figures 2 and 3.

Since cellulosic paper artifacts that need cleaning and restoration are usually written, drawn or printed on, in conjunction with their cleaning effect, other important aspects of ILs are related to their maintaining intact the respective addition drawing/text during the cleaning process and that the ILs mobilize preferentially only the degraded compounds, not the inks, pigments or toners.

The spectrophotometric VIS analysis of the pure ILs in which the ink-dyed Whatman filter paper strips were immersed for 24 h (Figure 6) have shown that all of the ILs have the ability to solubilize a low amount of the iron gall ink. Even if this leached amount is very low and could actually represent the excess of sorbed ink on the surface of the paper, while visually the strips remain basically unchanged, the exact impact on a real-life artifact needs to be further assessed.

The highest solvation ability for the iron gall ink is attributed to the lowest molar mass IL—PrEIMTs—while for the longer alkyl-chain ones, the leached amounts are close to the detection limit of the instrument. To evaluate the compatibility of the ILs with handwritten supports, a strip of B1 paper was inscribed with various lines and word-like shapes using iron gall ink and a fountain pen with a O type calligraphy nib. After drying for 24 h, the strip was immersed in the HeptMIMTs 10% v/v isopropanol solution for 6 h and then extracted, dried and visually inspected. HeptMIMTs is an IL that has a long alkyl chain cation and a weakly basic anion, which can reduce the solvency and reactivity of the IL towards cellulose and iron gall ink, respectively.

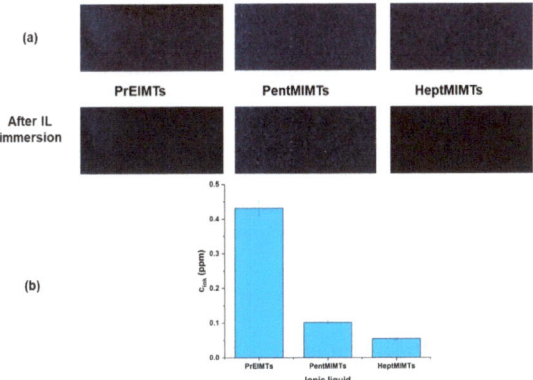

Figure 6. (a) Photographic images of the dyed Whatman paper strips immersed in ILs; (b) concentration of leached ink in the ILs.

Figure 7 shows that the iron gall ink drawings remained intact and unaffected after the IL treatment, indicating that HeptMIMTs did not compromise or erase the integrity of the ink, while the paper itself became lighter and cleaner after the IL treatment.

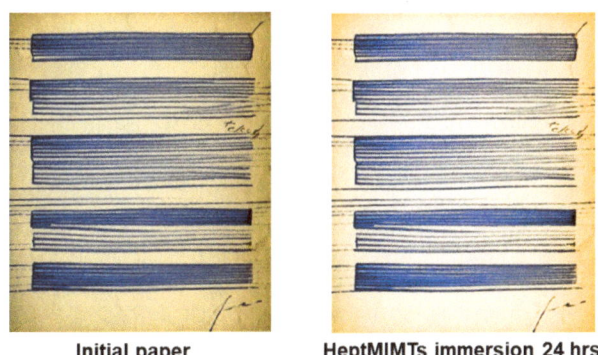

Figure 7. Photographic images of the inked B1 paper supports before and after 6 h immersion in the HeptMIMTs 10% v/v isopropanol solution.

3. Materials and Methods

The ionic liquids employed in this study—1-ethyl-3-propylimidazolium bis(trifluoromethylsulfonyl)imide ($C_{10}H_{15}F_6N_3O_4S_2$), 1-methyl-3-pentylimidazolium bis(trifluoromethylsulfonyl)imide ($C_{11}H_{17}F_6N_3O_4S_2$), and 1-heptyl-3-methylimidazolium bis(trifluoromethylsulfonyl)imide ($C_{13}H_{21}F_6N_3O_4S_2$)—were sourced from IoLiTec Ionic Liquids Technologies GmbH, Heilbronn, Germany, with a purity of 98% or greater (Table 2). Isopropyl alcohol (Sigma-Aldrich, Darmstadt, Germany, >99% vol.) was used as a basis to prepare 10% v/v and 50% v/v mixtures with the three ILs.

Paper strips were extracted from a book published in 1938 (B1). Several characteristics of this paper, such as its basis weight, pH and basic composition are given in Table 2. A Whatman qualitative filter Grade 1, with α-cellulose content surpassing 98%, served as a reference for comparative analysis.

Before immersion in the reconditioning solutions and pure ionic liquids, all the paper strips underwent a conditioning period of 24 h at 23 °C ± 1 °C and 50 ± 2% relative humidity within a desiccator containing a supersaturated $Mg(NO_3)_2$ solution. The initial weight of the paper strips was measured (m_0) to establish a baseline for comparative evaluations.

The Whatman filter paper strips underwent immersion in pure ionic liquids (ILs) as well as in 10% v/v and 50% v/v IL/isopropyl alcohol mixtures. The strips were periodically removed and weighed until reaching a state of constant mass, indicating equilibrium sorption. This equilibration process took approximately 6 h from the start of immersion. Subsequent to achieving equilibrium, the samples underwent conditioning by following the previously outlined procedure and were reweighed (m_f).

Since isopropyl alcohol is volatile enough to undergo complete evaporation, the paper strips in this scenario retained only the ionic liquids. The specific consumption of IL, reported per gram of paper (S_c), was calculated using Equation (1) [23]:

$$S_c = \frac{m_f - m_0}{m_0} \quad (1)$$

For the treatment of B1 paper strips, a 10% v/v solution of ionic liquids (ILs) in isopropyl alcohol was specifically selected, coupled with a 6 h treatment duration. This choice was made to facilitate the rapid absorption of the solution into the paper strips, owing to its lower viscosity and the minimal IL content.

The ILs used in this study have the same anion—bis(trifluoromethylsulfonyl) imide—which can form weak hydrogen bonds with cellulose and small-molecular compounds [59–61]. They also have similar cations, which are N-alkyl-N'-methylimidazolium derivatives with different alkyl chain lengths. These cations have acidic protons on the imidazolium ring that can form C-H···O hydrogen bonds with the compounds found in paper, which partly explains their cleaning effect and deacidification capabilities. However, the cations also have side alkyl chains that can reduce the solubility of cellulose by steric hindrance or hydrophobic interactions [60], limiting potential damage in uncontrolled swelling or dissolution of cellulose fibers. Therefore, the interaction of these ILs with cellulose depend on the balance between the anion and the cation effects, as well as the length of the alkyl chain. The ILs from this study were chosen because in our preliminary screening studies they did not impart modifications to paper texture, in contrast to short alkyl chain ILs, such as 1-ethyl-3-methylimidazolium or 1-butyl-3-methylimidazolium with chloride or acetate anions, reported in the literature as solvents for cellulose [28].

To safeguard valuable cellulosic artifacts, extend their longevity, and mitigate potential long-term damage—coupled with considerations of economic feasibility—the decision was made to exclusively employ the 10% v/v concentration in the treatment of B1 paper samples.

Table 2. Characteristics of the used ILs and B1 paper.

Ionic Liquids Used [1]	B1 Paper Characteristics
PrEIMTs (391), cation with $(CF_3SO_2)_2N^-$ anion	pH [2]: 4.30 ± 0.15
PentMIMTs (405), cation with $(CF_3SO_2)_2N^-$ anion	Paper basis weight: 67 g/m²
HeptMIMTs (433), cation with $(CF_3SO_2)_2N^-$ anion	Composition [3]: 57% groundwood fiber content, 43% bleached chemical pulp, 1.1% ash content

[1] Molar masses of ILs are given in parenthesis; [2] the pH values of paper were determined in accordance with Test Method TAPPI/ANSI T 529 om-21 [62]; [3] determined according to [63] (fiber and chemical pulp content) and [64] (ash content), average of triplicate measurements.

The B1 paper strips underwent an identical treatment and conditioning process as the Whatman filter paper. Subsequent to the treatment, the pH of the paper was measured following the T 529-om-21 protocol [62]. This measurement served to assess the efficacy of ionic liquids (ILs) in deacidifying the paper.

The IL-treated paper samples were coded as B1-PrEIMTs, B1-PentMIMTs, and B1-HeptMIMTs throughout this work.

To evaluate the potential advantages of ionic liquids (ILs) in the cleaning process of paper, color modifications in the B1 paper substrates before and after IL treatment were analyzed using a PCEXXM30 colorimeter (PCE Instruments, Meschede, Germany), operating within the CIE-Lab* color space. In this color space, L* signifies lightness, while a* and b* denote the red/green and yellow/blue coordinates, respectively. The IL-treated B1 strips underwent an additional accelerated UV weathering test at 23 °C and 54% relative humidity within a Bio-Link model BLX-E254 (Fisher Scientific, Wien, Austria) irradiation chamber, exposed to 254 nm wavelength for 10 h at a total energy density of 8 mJ/cm^2. Post-irradiation, color parameters were re-measured to assess any changes and gauge the potential efficacy of ILs as UV stabilizers. This approach aimed to quantify both the immediate impact of IL treatment on paper color and its potential role in providing protection against UV-induced color alterations.

To calculate the total color differences (ΔE^*) for each sample, the following equation (Equation (2)) was utilized [65]. The subscripts "f" and "i" correspond to the values of the color parameters after UV aging and IL treatment, and before, respectively:

$$\Delta E^* = \sqrt{\left(L_f^* - L_i^*\right)^2 + \left(a_f^* - a_i^*\right)^2 + \left(b_f^* - b_i^*\right)^2} \qquad (2)$$

The optical microscopy images were acquired with a Leica DM_ILM microscope (Leica Microsystems, Wetzlar, Germany).

ATR-FTIR spectra of the ILs, B1 paper and IL-treated B1 paper were acquired using a Perkin-Elmer Spectrum BXII spectrometer (Waltham, MA, USA) in the spectral range of 650–4000 cm^{-1}. The spectra were obtained with a scan step of 2 cm^{-1}, and 10 averaged scans were recorded for each sample.

The impact of ionic liquids on iron gall ink, a specific ink commonly associated with paper artifacts, was investigated in this study. Whatman filter paper strips were immersed for 15 min in an iron gall ink formulation (KWZ Iron Gall Ink, Blue-Black, Archive Iron Gall type, Warsaw, Poland). Subsequently, the dyed paper strips were clipped on string supports and allowed to vertically air-dry for 24 h before initiating ink stability testing.

For the evaluation of ink stability, the dyed paper strips were immersed in pure ionic liquids for 24 h at room temperature (22 °C). Following this, the ionic liquids underwent spectrophotometric analysis to ascertain whether the ink leached from the paper supports. The spectrophotometric analysis, conducted on a UV-VIS TU1801 spectrophotometer (DIY-power Co., Ltd., Guangzhou, China) within the range of 400 to 750 nm, utilized the neat ionic liquid as a background reference. Several ink solutions in ILs were prepared, in concentrations ranging from 2 to 12 ppm. Notably, Figure 8a (for the maximum 12 ppm concentration) illustrates that the ink exhibits a broad absorption maximum, dependent on the type of ionic liquid, spanning from 565 nm (for PrEIMTs) to 600 nm for PentMIMTs.

Calibration curves, depicted in Figure 8b, were constructed for the ink dissolved in each respective ionic liquid in a concentration range from 2 to 12 ppm. These curves served as a quantitative tool to evaluate the extent of paper de-inking achieved by each IL. The ILs in which the dyed Whatman paper strips had been immersed underwent subsequent spectrophotometric analysis with neat ILs as background, to quantitatively assess the de-inking of the paper supports.

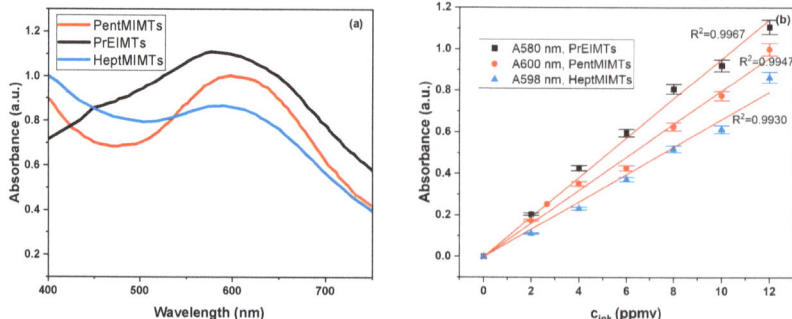

Figure 8. (**a**) VIS spectra of the iron gall ink in the three ILs at 12 ppm concentration; (**b**) the calibration curves for the ink dissolved in the three ILs.

4. Conclusions

In conclusion, this study underscores the significant potential of ionic liquids (ILs) in the restoration of aged paper artifacts, with specific findings highlighting the prominent deacidification capabilities of ILs with the 1-heptyl-3-methylimidazolium cation, leading to a neutral pH. The observed color enhancements, particularly with 1-heptyl-3-methylimidazolium and 1-pentyl-3-methylimidazolium ILs, present promising prospects for revitalizing aged or degraded paper. Furthermore, the investigation into ILs as UV stabilizers for paper reveals encouraging outcomes, emphasizing the role of 1-heptyl-3-methylimidazolium IL in minimizing yellowing post-UV irradiation.

This research contributes to the broader field of paper conservation and cultural heritage preservation, offering practical insights into the application of ILs for restoring and maintaining paper-based artifacts. The results highlight the nuanced effects of different ILs on paper properties, providing useful information for conservators and researchers.

Future studies could delve deeper into the long-term effects of IL treatment on paper artifacts and explore additional IL formulations. The preservation of inked materials and a comprehensive understanding of ILs' interaction with diverse inks warrant further investigation. Additionally, exploring the scalability and practical implementation of IL-based restoration processes would enhance the applicability of these findings in real-world conservation practices. This work paves the way for a nuanced understanding of ILs' potential in cultural heritage preservation, opening avenues for sustainable and effective strategies in the field of paper restoration.

Author Contributions: Conceptualization, C.C.; methodology, C.C.; software, I.C.R.; validation, C.C. and I.C.R.; formal analysis, C.C.; investigation, C.C. and I.C.R.; resources, C.C.; data curation, C.C.; writing—original draft preparation, C.C. and I.C.R.; writing—review and editing, C.C.; visualization, I.C.R.; supervision, C.C.; project administration, C.C.; funding acquisition, C.C. All authors have read and agreed to the published version of the manuscript.

Funding: This research received no external funding.

Institutional Review Board Statement: Not applicable.

Informed Consent Statement: Not applicable.

Data Availability Statement: The data presented in this study are available on request from the corresponding author.

Conflicts of Interest: The authors declare no conflicts of interest.

References

1. Marijnissen, R. Degradation, Conservation, and Restoration of Works of Art: Historical Overview. *CeROArt* **2015**, *6*, 122–135. [CrossRef]
2. Chen, T.; Su, R.; Zhang, Y.; Zhang, J.; Yu, H. The Smart Conservation System of Ancient Books Driven by Smart Data. *J. Libr. Sci. China* **2023**, *49*, 68–81. [CrossRef]
3. Stiglitz, M. Historical Perspectives in the Conservation of Works of Art on Paper. *J. Inst. Conserv.* **2017**, *40*, 83–84. [CrossRef]
4. Strlič, M.; Thomas, J.; Trafela, T.; Cséfalvayová, L.; Cigić, I.K.; Kolar, J.; Cassar, M. Material Degradomics: On the Smell of Old Books. *Anal. Chem.* **2009**, *81*, 8617–8622. [CrossRef]
5. Chiriu, D.; Ricci, P.C.; Cappellini, G.; Salis, M.; Loddo, G.; Carbonaro, C.M. Ageing of Ancient Paper: A Kinetic Model of Cellulose Degradation from Raman Spectra. *J. Raman Spectrosc.* **2018**, *49*, 1802–1811. [CrossRef]
6. Hamburger, S. Preservation and Conservation for Libraries and Archives. *Libr. Collect. Acquis. Tech. Serv.* **2005**, *29*, 444–445. [CrossRef]
7. Ahn, K.; Rosenau, T.; Potthast, A. The Influence of Alkaline Reserve on the Aging Behavior of Book Papers. *Cellulose* **2013**, *20*, 1989–2001. [CrossRef]
8. Ahn, K.; Henniges, U.; Banik, G.; Potthast, A. Is Cellulose Degradation Due to β-Elimination Processes a Threat in Mass Deacidification of Library Books? *Cellulose* **2012**, *19*, 1149–1159. [CrossRef]
9. Joseph, E. *Microorganisms in the Deterioration and Preservation of Cultural Heritage*; Springer: Cham, Switzerland, 2021.
10. Kačík, F.; Kačíková, D.; Jablonský, M.; Katuščák, S. Cellulose Degradation in Newsprint Paper Ageing. *Polym. Degrad. Stab.* **2009**, *94*, 1509–1514. [CrossRef]
11. Potthast, A.; Henniges, U.; Banik, G. Iron Gall Ink-Induced Corrosion of Cellulose: Aging, Degradation and Stabilization. Part 1: Model Paper Studies. *Cellulose* **2008**, *15*, 849–859. [CrossRef]
12. Dubois, A. *Solving Cases: Book and Paper Artefact Restoration*; Scriptorium: Turnhout, Belgium, 2013; Volume 67.
13. Emanuele, L.; Dujaković, T.; Roselli, G.; Campanelli, S.; Bellesi, G. The Use of a Natural Polysaccharide as a Solidifying Agent and Color-Fixing Agent on Modern Paper and Historical Materials. *Organics* **2023**, *4*, 21. [CrossRef]
14. Bicchieri, M.; Monti, M.; Antonelli, M.L. A New Low-Cost and Complete Restoration Method: A Simultaneous Non-Aqueous Treatment of Deacidification and Reduction. In Proceedings of the Third International Conference on Science and Technology for the Safeguard of Cultural Heritage in the Mediterranean Basin, Alcalá de Henares, Spain, 9–14 July 2001.
15. Henniges, U.; Potthast, A. Bleaching Revisited: Impact of Oxidative and Reductive Bleaching Treatments on Cellulose and Paper. *Restaurator* **2009**, *30*, 294–320. [CrossRef]
16. Mohie, M.A.; Ismail, S.A.; Hassan, A.A.; Tawfik, A.M.; Mohamed, W.S. Assessment of the Applicability of Cellulolytic Enzyme in Disassembling of Caked Papers. *Egypt. J. Chem.* **2022**, *65*, 581–591. [CrossRef]
17. Adamo, M.; Magaudda, G.; Tata, A. Radiation Technology for Cultural Heritage Restoration. *Restaurator* **2004**, *25*, 159–170. [CrossRef]
18. Fotakis, C.; Kautek, W.; Castillejo, M. Lasers in the Preservation of Cultural Heritage. *Laser Chem.* **2006**, *2006*, 074791. [CrossRef]
19. Eastaugh, N.; Needles, H.L.; Zeronian, S.H.; Zeronian, S.H.; Needles, H.L. Historic Textile and Paper Materials: Conservation and Characterization. *Stud. Conserv.* **1990**, *35*, 231. [CrossRef]
20. Ahn, K.; Banik, G.; Potthast, A. Sustainability of Mass-Deacidification. Part II: Evaluation of Alkaline Reserve. *Restaurator* **2012**, *33*, 48–75. [CrossRef]
21. Potthast, A.; Ahn, K. Critical Evaluation of Approaches toward Mass Deacidification of Paper by Dispersed Particles. *Cellulose* **2017**, *24*, 323–332. [CrossRef]
22. Keraite, G.; Sivakova, B.; Kiuberis, J. Investigation of the Impact of Organic and Inorganic Halides on the Ageing Stability of Paper with Iron Gall Ink. *Chemija* **2017**, *28*, 137–147.
23. Kozirog, A.; Wysocka-Robak, A. Application of Ionic Liquids in Paper Properties and Preservation. In *Progress and Developments in Ionic Liquids*; IntechOpen: London, UK, 2017.
24. Santos, S.M.; Carbajo, J.M.; Quintana, E.; Ibarra, D.; Gomez, N.; Ladero, M.; Eugenio, M.E.; Villar, J.C. Characterization of Purified Bacterial Cellulose Focused on Its Use on Paper Restoration. *Carbohydr. Polym.* **2015**, *116*, 173–181. [CrossRef]
25. Gueidão, M.; Vieira, E.; Bordalo, R.; Moreira, P. Available Green Conservation Methodologies for the Cleaning of Cultural Heritage: An Overview. *Estud. Conserv. Restauro* **2020**, *12*, 22–44. [CrossRef]
26. Baglioni, M.; Poggi, G.; Chelazzi, D.; Baglioni, P. Advanced Materials in Cultural Heritage Conservation. *Molecules* **2021**, *26*, 3967. [CrossRef] [PubMed]
27. Pernak, J.; Jankowska, N.; Walkiewicz, F.; Jankowska, A. The Use of Ionic Liquids in Strategies for Saving and Preserving Cultural Artifacts. *Pol. J. Chem.* **2008**, *82*, 2227–2230.
28. Olsson, C.; Hedlund, A.; Idström, A.; Westman, G. Effect of Methylimidazole on Cellulose/Ionic Liquid Solutions and Regenerated Material Therefrom. *J. Mater. Sci.* **2014**, *49*, 3423–3433. [CrossRef]
29. Shamsuri, A.A.; Abdan, K.; Kaneko, T. A Concise Review on the Physicochemical Properties of Biopolymer Blends Prepared in Ionic Liquids. *Molecules* **2021**, *26*, 216. [CrossRef] [PubMed]
30. Caminiti, R.; Campanella, L.; Plattner, S.H.; Scarpellini, E. Effects of Innovative Green Chemical Treatments on Paper. Can They Help in Preservation? *Int. J. Conserv. Sci.* **2016**, *7*, 247–258.

31. Schmitz, K.; Wagner, S.; Reppke, M.; Maier, C.L.; Windeisen-Holzhauser, E.; Philipp Benz, J. Preserving Cultural Heritage: Analyzing the Antifungal Potential of Ionic Liquids Tested in Paper Restoration. *PLoS ONE* **2019**, *14*, e0219650. [CrossRef]
32. Przybysz, K.; Drzewińska, E.; Stanisławska, A.; Wysocka-Robak, A.; Cieniecka-Rosłonkiewicz, A.; Foksowicz-Flaczyk, J.; Pernak, J. Ionic Liquids and Paper. *Ind. Eng. Chem. Res.* **2005**, *44*, 4599–4604. [CrossRef]
33. Ocreto, J.B.; Chen, W.H.; Rollon, A.P.; Chyuan Ong, H.; Pétrissans, A.; Pétrissans, M.; De Luna, M.D.G. Ionic Liquid Dissolution Utilized for Biomass Conversion into Biofuels, Value-Added Chemicals and Advanced Materials: A Comprehensive Review. *Chem. Eng. J.* **2022**, *445*, 136733. [CrossRef]
34. Croitoru, C.; Patachia, S.; Porzsolt, A.; Friedrich, C. Effect of Alkylimidazolium Based Ionic Liquids on the Structure of UV-Irradiated Cellulose. *Cellulose* **2011**, *18*, 1469–1479. [CrossRef]
35. Helbrecht, C.; Schmitt, F.; Meckel, T.; Biesalski, M.; Etzold, B.J.M.; Schabel, S. Mechanical Properties of Paper Saturated With a Hydrophobic Ionic Liquid. *Bioresources* **2023**, *18*, 2842–2856. [CrossRef]
36. Croitoru, C.; Roata, I.C. Ionic Liquids as Potential Cleaning and Restoration Agents for Cellulosic Artefacts. *Processes* **2024**, *12*, 341. [CrossRef]
37. Dimitrić, N.; Spremo, N.; Vraneš, M.; Belić, S.; Karaman, M.; Kovačević, S.; Karadžić, M.; Podunavac-Kuzmanović, S.; Korolija-Crkvenjakov, D.; Gadžurić, S. New Protic Ionic Liquids for Fungi and Bacteria Removal from Paper Heritage Artefacts. *RSC Adv.* **2019**, *9*, 17905–17912. [CrossRef] [PubMed]
38. Marín, E.; Sistach, M.C.; Jiménez, J.; Clemente, M.; Garcia, G.; García, J.F. Distribution of Acidity and Alkalinity on Degraded Manuscripts Containing Iron Gall Ink. *Restaurator* **2015**, *36*, 229–247. [CrossRef]
39. Mochizuki, Y.; Itsumura, H.; Enomae, T. Mechanism of Acidification That Progresses in Library Collections of Books Made of Alkaline Paper. *Restaurator. Int. J. Preserv. Libr. Arch. Mater.* **2020**, *41*, 153–172. [CrossRef]
40. Wójciak, A. Washing, Spraying and Brushing. A Comparison of Paper Deacidification by Magnesium Hydroxide Nanoparticles. *Restaurator. Int. J. Preserv. Libr. Arch. Mater.* **2015**, *36*, 3–23. [CrossRef]
41. Anbardan, S.Z.; Mokhtari, J.; Yari, A.; Bozcheloei, A.H. Direct Synthesis of Amides and Imines by Dehydrogenative Homo or Cross-Coupling of Amines and Alcohols Catalyzed by Cu-MOF. *RSC Adv.* **2021**, *11*, 20788–20793. [CrossRef] [PubMed]
42. Vujcic, I.; Masic, S.; Medic, M.; Milicevic, B.; Dramicanin, M. The Influence of Gamma Irradiation on the Color Change of Wool, Linen, Silk, and Cotton Fabrics Used in Cultural Heritage Artifacts. *Radiat. Phys. Chem.* **2019**, *156*, 307–313. [CrossRef]
43. Luo, Y.; Xiang, Y.; Yang, Q.; Liu, J. Characterization of UVA-Irradiated Wheat Paste and Paste-Coated Paper. *J. Cult. Herit.* **2023**, *64*, 150–159. [CrossRef]
44. Croitoru, C.; Roata, I.C. Ionic Liquids as Antifungal Agents for Wood Preservation. *Molecules* **2020**, *25*, 4289. [CrossRef]
45. Patachia, S.; Croitoru, C.; Friedrich, C. Effect of UV Exposure on the Surface Chemistry of Wood Veneers Treated with Ionic Liquids. *Appl. Surf. Sci.* **2012**, *258*, 6723–6729. [CrossRef]
46. Zuo, Y.; Lv, J.; Wei, N.; Chen, X.; Tong, J. Effect of Anions and Cations on the Self-Assembly of Ionic Liquid Surfactants in Aqueous Solution. *J. Mol. Liq.* **2023**, *375*, 121342. [CrossRef]
47. Motamedian, H.R.; Halilovic, A.E.; Kulachenko, A. Mechanisms of Strength and Stiffness Improvement of Paper after PFI Refining with a Focus on the Effect of Fines. *Cellulose* **2019**, *26*, 4099–4124. [CrossRef]
48. Zervos, S.; Alexopoulou, I. Paper Conservation Methods: A Literature Review. *Cellulose* **2015**, *22*, 2859–2897. [CrossRef]
49. Zaitsau, D.H.; Abdelaziz, A. The Study of Decomposition of 1-Ethyl-3-Methyl-Imidazolium Bis(Trifluoromethylsulfonyl)Imide by Using Termogravimetry: Dissecting Vaporization and Decomposition of ILs. *J. Mol. Liq.* **2020**, *313*, 113507. [CrossRef]
50. Kiefer, J.; Fries, J.; Leipertz, A. Experimental Vibrational Study of Imidazolium-Based Ionic Liquids: Raman and Infrared Spectra of 1-Ethyl-3-Methylimidazolium Bis(Trifluoromethylsulfonyl)Imide and 1-Ethyl-3-Methylimidazolium Ethylsulfate. *Appl. Spectrosc.* **2007**, *61*, 1306–1311. [CrossRef] [PubMed]
51. Heimer, N.E.; Del Sesto, R.E.; Meng, Z.; Wilkes, J.S.; Carper, W.R. Vibrational Spectra of Imidazolium Tetrafluoroborate Ionic Liquids. *J. Mol. Liq.* **2006**, *124*, 84–95. [CrossRef]
52. Khachatrian, A.A.; Rakipov, I.T.; Mukhametzyanov, T.A.; Solomonov, B.N.; Miroshnichenko, E.A. The Ability of Ionic Liquids to Form Hydrogen Bonds with Organic Solutes Evaluated by Different Experimental Techniques. Part II. Alkyl Substituted Pyrrolidinium- and Imidazolium-Based Ionic Liquids. *J. Mol. Liq.* **2020**, *309*, 113138. [CrossRef]
53. Hospodarova, V.; Singovszka, E.; Stevulova, N. Characterization of Cellulosic Fibers by FTIR Spectroscopy for Their Further Implementation to Building Materials. *Am. J. Anal. Chem.* **2018**, *9*, 303–310. [CrossRef]
54. Younis, O.M.; El Hadidi, N.M.N.; Darwish, S.S.; Mohamed, M.F. Preliminary Study on the Strength Enhancement of Klucel E with Cellulose Nanofibrils (CNFs) for the Conservation of Wooden Artifacts. *J. Cult. Herit.* **2023**, *60*, 41–49. [CrossRef]
55. Hong, T.; Yin, J.Y.; Nie, S.P.; Xie, M.Y. Applications of Infrared Spectroscopy in Polysaccharide Structural Analysis: Progress, Challenge and Perspective. *Food Chem. X* **2021**, *12*, 100168. [CrossRef] [PubMed]
56. Xu, K.; Wang, J. Discovering the Effect of Alum on UV Photo-Degradation of Gelatin Binder via FTIR, XPS and DFT Calculation. *Microchem. J.* **2019**, *149*, 103934. [CrossRef]
57. Littlejohn, D.; Pethrick, R.A.; Quye, A.; Ballany, J.M. Investigation of the Degradation of Cellulose Acetate Museum Artefacts. *Polym. Degrad. Stab.* **2013**, *98*, 416–424. [CrossRef]
58. Erukhimovitch, V.; Pavlov, V.; Talyshinsky, M.; Souprun, Y.; Huleihel, M. FTIR Microscopy as a Method for Identification of Bacterial and Fungal Infections. *J. Pharm. Biomed. Anal.* **2005**, *37*, 1105–1108. [CrossRef] [PubMed]

59. Zhou, Y.; Zhang, X.; Yin, D.; Zhang, J.; Mi, Q.; Lu, H.; Liang, D.; Zhang, J. The Solution State and Dissolution Process of Cellulose in Ionic-Liquid-Based Solvents with Different Hydrogen-Bonding Basicity and Microstructures. *Green Chem.* **2022**, *24*, 3824–3833. [CrossRef]
60. Lu, B.; Xu, A.; Wang, J. Cation Does Matter: How Cationic Structure Affects the Dissolution of Cellulose in Ionic Liquids. *Green Chem.* **2014**, *16*, 1326–1335. [CrossRef]
61. Nag, N.; Sharma, C.; Singh, A.; Roy, B.N.; Sharma, S.K.; Kumar, A. Trifluorosulfonyl Imide-Based Ionic Liquid Electrolytes for Lithium-Ion Battery: A Review. *J. Inst. Eng. (India) Ser. D* **2023**, *104*, 427–436. [CrossRef]
62. *T 509 om-02*; Hydrogen Ion Concentration (pH) of Paper Extracts (Cold Extraction Method). TAPPI: Atlanta, GA, USA, 2002.
63. Malešič, J.; Kraševec, I.; Cigić, I.K. Determination of Cellulose Degree of Polymerization in Historical Papers with High Lignin Content. *Polymer* **2021**, *13*, 1990. [CrossRef]
64. Małachowska, E.; Dubowik, M.; Boruszewski, P.; Łojewska, J.; Przybysz, P. Influence of Lignin Content in Cellulose Pulp on Paper Durability. *Sci. Rep.* **2020**, *10*, 19998. [CrossRef]
65. Croitoru, C.; Patachia, S.; Doroftei, F.; Parparita, E.; Vasile, C. Ionic Liquids Influence on the Surface Properties of Electron Beam Irradiated Wood. *Appl. Surf. Sci.* **2014**, *314*, 956–966. [CrossRef]

Disclaimer/Publisher's Note: The statements, opinions and data contained in all publications are solely those of the individual author(s) and contributor(s) and not of MDPI and/or the editor(s). MDPI and/or the editor(s) disclaim responsibility for any injury to people or property resulting from any ideas, methods, instructions or products referred to in the content.

Article

Anti-Cracking TEOS-Based Hybrid Materials as Reinforcement Agents for Paper Relics

Mengruo Wu [1], Le Mu [2], Zhiyue Zhang [1], Xiangna Han [1,*], Hong Guo [1,*] and Liuyang Han [1]

[1] Key Laboratory of Archaeomaterials and Conservation, Ministry of Education, Institute for Cultural Heritage and History of Science & Technology, University of Science and Technology Beijing, Beijing 100083, China; d202310767@xs.ustb.edu.cn (M.W.); zhangzhiyue98@163.com (Z.Z.); hanliuyang@ustb.edu.cn (L.H.)
[2] Baotou Museum, Baotou 014010, China; mule13948421468@126.com
* Correspondence: jayna422@ustb.edu.cn (X.H.); guohong@ustb.edu.cn (H.G.)

Abstract: Tetraethoxysilane (TEOS) is the most commonly used silicon-based reinforcement agent for conserving art relics due to its cost effectiveness and commercial maturity. However, the resulting silica gel phase is prone to developing cracks as the gel shrinks during the sol–gel process, potentially causing severe damage to the objects being treated. In this study, dodecyltrimethoxysilane (DTMS) was introduced into TEOS to minimize this shrinkage by adding elastic long chains to weaken the capillary forces. The gel formed from the DTMS/TEOS hybrid material was transparent and crack-free, featuring a dense microstructure without mesopores or micropores. It exhibited excellent thermal stability, with a glass transition temperature of up to 109.64 °C. Evaluation experiments were conducted on artificially aged, handmade bamboo paper. The TEOS-based hybrid material effectively combined with the paper fibers through the sol–gel process, polymerizing into a network structure that enveloped the paper surface or penetrated between the fibers. The surface of the treated paper displayed excellent hydrophobic properties, with no significant changes in appearance, color, or air permeability. The mechanical properties of the treated bamboo paper improved significantly, with longitudinal and transverse tensile strengths increasing by up to 36.63% and 44.25%, respectively. These research findings demonstrate the promising potential for the application of DTMS/TEOS hybrid materials in reinforcing paper relics.

Keywords: paper relics; organosilane; reinforcement; preservation

Citation: Wu, M.; Mu, L.; Zhang, Z.; Han, X.; Guo, H.; Han, L. Anti-Cracking TEOS-Based Hybrid Materials as Reinforcement Agents for Paper Relics. *Molecules* **2024**, *29*, 1834. https://doi.org/10.3390/molecules29081834

Academic Editors: Yueer Yan, Yi Tang and Yuliang Yang

Received: 20 March 2024
Revised: 13 April 2024
Accepted: 16 April 2024
Published: 17 April 2024

Copyright: © 2024 by the authors. Licensee MDPI, Basel, Switzerland. This article is an open access article distributed under the terms and conditions of the Creative Commons Attribution (CC BY) license (https://creativecommons.org/licenses/by/4.0/).

1. Introduction

Organosilane materials have shown potential as ideal paper relic preservation materials due to their dual advantages of both organic and inorganic properties. Characterized by Si–O–Si bonds as the primary skeleton, these materials possess thermal and chemical stability far superior to general polymer materials, resembling the properties of inorganic materials. During the sol–gel process, organosilane will react with the hydroxyl groups upon cellulose condensation, forming Si–O–C bonds that can effectively reinforce the paper. Furthermore, organosilane materials contain Si–C bonds with diverse organic groups, enabling the introduction of various groups to achieve deacidification [1], hydrophobicity [2], antimicrobial [3], and other desired effects.

Currently, silane coupling agents have been involved in the widespread practice of paper heritage protection. These agents belong to a class of reactive silanes that possess organic functional groups. Some of these compounds carry organic groups, such as alkyl, vinyl, amino, epoxy, or sulfhydryl groups, which enhance reactivity with organic compounds and introduce various functionalities into the resulting polymers. However, some silanes contain inorganic groups, such as alkoxy, aryloxy, acyl, and chlorine, which have excellent reactivity with the surfaces of inorganic substances [4–6]. When applied to paper, silane will undergo hydrolysis to produce silanols. These silanols will then dehydrate, condense, and crosslink to form oligosiloxanes, and the silanol groups present in oligosiloxanes will adhere to the paper's surface [7]. Using the silane coupling method enables the grafting and modification of paper cellulose [8], introducing various functional groups into the paper and thereby endowing it with a wide range of appealing properties. It has been reported that silane coupling agents could bestow surface hydrophobicity and water resistance upon the paper [9,10] and even transform it into a catalyst medium or a drug carrier [11–13].

The application of silane coupling agents has also been explored in the preservation of paper relics. Chen et al. [14] and Shen et al. [15] both employed three silane compounds, namely, N-aminoethyl-3-aminopropyl-triethoxysilane (AETAPTES, trademarked as KH791), 3-aminopropyl-diethoxymethylsilane (AMDES, trademarked as JH-M902), and 3-aminopropyl-triethoxysilane (APTES, trademarked as KH550), for the reinforcement treatment of Xuan paper. These agents were dissolved in anhydrous ethanol to form a reinforcing solution. When the paper was immersed in this solution, the studies found that all three silanes effectively reinforced the paper, enhancing its tensile strength and folding durability. Among these, AETAPTES demonstrated the most notable reinforcement effect. Specifically, when applied at a concentration of 15%, the solutions increased the tensile strength of the paper by approximately threefold, and the folding endurance was 60 times greater.

Since 2006, researchers from the Muséum National d'Histoire Naturelle in France have conducted a comprehensive series of studies focusing primarily on the utilization of amino-functionalized silanes for paper conservation [1,16–20]. For the main purpose of deacidification, these studies evaluated various silanes, including 3-aminopropyltrimethoxysilane (ATMS), 3-aminopropyltriethoxysilane (ATES, identical to APTES), 4-amino-3,3-dimethylbutyltrimethoxysilane (ADBTMS), hexamethyl disiloxane (HMDS), and dimethylaminopropylmethyl diethoxysilane (DMAPDES), as well as other silane coupling agents, along with their corresponding compounds. These agents were tested for their efficacy in protecting both naturally aged and artificially aged paper. The results revealed that these silane coupling agents, apart from their deacidifying capabilities, could significantly enhance the mechanical properties of paper, such as tensile strength and folding endurance, by forming a robust network structure on the fiber surfaces.

Current studies have established that organosilane materials exhibit favorable reinforcing effects on paper relics; however, most of the silane coupling agents used either are costly or rely on laboratory synthesis. Considering the need for a substantial quantity of materials to support the preservation of paper relics, it remains imperative to explore cost-effective and readily available alternatives. Tetraethoxysilane (TEOS), a commercially viable and economical organosilane material, is most frequently utilized in heritage conservation [21] and offers a promising avenue if modifications can be made to TEOS to align with the performance requirements of paper relic protection.

The primary challenge with TEOS is its susceptibility to cracking upon gelation. This cracking can be attributed to two primary causes, namely, capillary stress during gel drying and the volatilization of water and alcohols during condensation [22]. Since the cracking of organosilanes may compromise the integrity of the artifact matrix, it has become imperative to engineer these materials with anti-cracking modifications. Although numerous strategies exist to address this issue, two primary methods remain particularly relevant for artifact preservation. First, the introduction of elastic chain segments may enhance the flexibility of the material, thereby resisting capillary stress during shrinkage [23]. For example, the incorporation of polydimethylsiloxane (PDMS) can densify the silicone gel structure [24–26]. In addition, the utilization of nanoparticles can increase the elastic modulus and gel pore size of the protective material, effectively mitigating capillary stress. This approach has been demonstrated through the use of nanoparticles such as TiO_2, SiO_2, and Al_2O_3 [27–29]. Furthermore, both methods can be combined to modify organosilanes. For example, Son et al. introduced nanometer-sized (3-glycidoxypropyl) trimethoxysilane (GPTMS) and polyhedral oligomeric silsesquioxane (POSS) with elastic chain segments into TEOS, resulting in a hybrid material providing a crack-free gel [30].

This study utilized inexpensive and widely available ethyl orthosilicate (TEOS) as the baseline material. To enhance its properties, dodecyltrimethoxysilane (DTMS) with flexible long chains was integrated into the TEOS agent, which resulted in a series of formations of DTMS/TEOS hybrid material agents through a compounding process. These DTMS/TEOS agents were evaluated based on their curing performance and were subsequently utilized to protect and treat aged bamboo paper. The color difference, mechanical properties, air penetration, and microstructure of the treated aged paper were characterized to assess the overall effectiveness of the DTMS/TEOS hybrid material for reinforcement and preservation.

2. Results and Discussion

2.1. DTMS/TEOS Hybrid Material

2.1.1. Curing Performance

In this study, we prepared a series of hybrid materials with varying mass ratios of MTDS and TEOS (W_{DTMS}/W_{TEOS} = 1:9, 2:8, 3:7, 4:6, and 5:5) and compared their curing effects. These five groups of materials were denoted as DT1 to DT5, respectively.

The appearance of the hybrid materials typically serves as a perceptive and fundamental criterion for assessing their curing performance. To ensure good curing and drying without cracking, the hybrid materials would need to be colorless and transparent to preserve the original appearance of the heritage samples during paper sample treatment. In addition, organosilanes undergo volume contraction during the sol–gel process, and excessive contraction can damage the paper fiber structure. Therefore, smaller volumetric shrinkage of the hybrid materials was sought.

Images of the DTMS/TEOS gels are shown in Figure 1a. As presented in Figure 1b, the appearance changes of the hybrid materials after gelation could be described as three typical cases. In this study, the cracks generated in the gels of all groups were confirmed as small cracks that did not cause the gels to become fragile, namely, fine cracks. The volume shrinkage of each formulation after gelation was different. For some groups, we observed that the shrinkage of the gel mainly occurred on the oxygen contact surface, and no significant impact was observed on the overall size and shape, which was defined as fine shrinkage. However, some gels with different formulations showed irregular shrinkage, which seriously affected the curing effect and was defined as severe shrinkage.

(a) Photographic images of the DTMS/TEOS hybrid material gels.

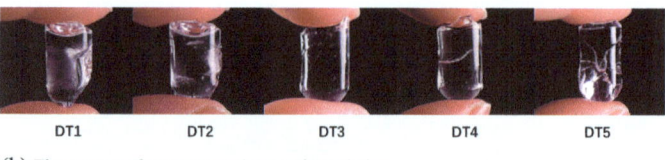

(b) Three types of appearance change after gelation.

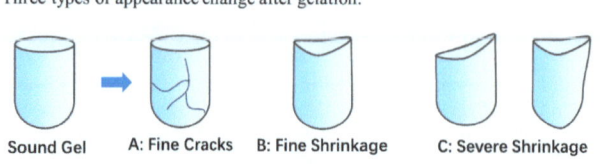

Figure 1. Appearance evaluation of the DTMS/TEOS hybrid material gels. (**a**) Photographic images of the DTMS/TEOS hybrid material gels; (**b**) three types of appearance change after gelation.

Table 1 presents information on the types of changes in appearance after gelation of each group, according to Figure 1b, as well as their volumetric shrinkage rates, mass retention rates, and curing times. The results indicated that under the catalytic conditions of 0.2 wt.% concentration dibutyltin dilaurate (DBTL), the DTMS/TEOS hybrid materials could meet the required curing performance. When the mass fraction of DTMS was only 10%, the system could produce a completely transparent gel with only fine cracks, and the volumetric shrinkage varied little among the five groups, mostly falling between 40% and 50%. When the mass fraction of DTMS was ≥40%, the gels exhibited low volumetric shrinkage, making them less prone to severe shrinkage and cracking. Among these, the DT5 agent (W_{DTMS}/W_{TEOS} = 5:5) demonstrated the most prominent anti-shrinkage effects, with the volumetric shrinkage of the gels reaching the lowest average value of only 36.72%.

Table 1. Appearance changes, mass retention rates, volumetric shrinkage rates, and curing times of the DTMS/TEOS hybrid materials.

Group	Appearance Change	Mass Retention Rate	Volumetric Shrinkage Rate	Curing Time
DT1	C	57.06%	42.11%	210~213 h
DT2	A C	55.51%	48.23%	
DT3	A	54.18%	50.12%	261 h
DT4	A B	59.77%	40.43%	
DT5	A	63.23%	36.72%	307 h

During the sol–gel process, organosilanes will release a significant amount of water and alcohol, with some organosilanes possessing a certain degree of volatility. Therefore, it was necessary to assess the retention of active ingredients by examining the mass retention rate of the hybrid materials. A higher retention rate typically indicates a greater coating effect on the paper. We found that the mass ratio of DTMS and TEOS had no significant impact on the mass retention rate of the DTMS/TEOS hybrid materials and generally remained within the range of 55–65%. Notably, the mass retention rate reached its highest point of 63.23% when W_{DTMS}/W_{TEOS} was set to 5:5.

Curing time also served as a crucial factor in determining whether the hybrid material agent could meet the required properties for practical applications. A longer curing time tended to enhance the stability of formulations during long-term practical applications, providing ample time for large-scale construction and fine processing. When the DTMS mass fraction increased, the full curing time also increased. The DT5 group, with the highest DTMS concentration, required the longest curing time at approximately 307 h. Taking both the curing time and cost into consideration, a formulation with a W_{DTMS}/W_{TEOS} ratio of 5:5 emerged as the most preferred option for maximizing active ingredient retention.

2.1.2. Thermal Stability

To investigate the thermal stability of the DTMS/TEOS hybrid materials, two groups with the largest and lowest W_{DTMS}/W_{TEOS} ratios (W_{DTMS}/W_{TEOS} = 1:9 and 5:5) were selected for DSC testing. The glass transition temperatures (Tg) of the hybrid materials were determined based on the DSC curves.

As shown in Figure 2, for the DTMS/TEOS hybrid materials, the Tg values were 80.16 °C and 109.64 °C when W_{DTMS}/W_{TEOS} = 1:9 and 5:5, respectively. Notably, all of the Tg temperatures were significantly higher than the temperature of a typical use environment (room temperature). These results indicated that the prepared hybrid materials would remain in a glassy state at the use environment temperature. In this state, the molecular chains and chain segment motions were effectively frozen, with only the atoms constituting molecular vibrations at their equilibrium positions. Consequently, the hybrid materials exhibited superior thermal stability performance [31], and due to their high Tg temperatures, these hybrid materials possessed greater rigidity.

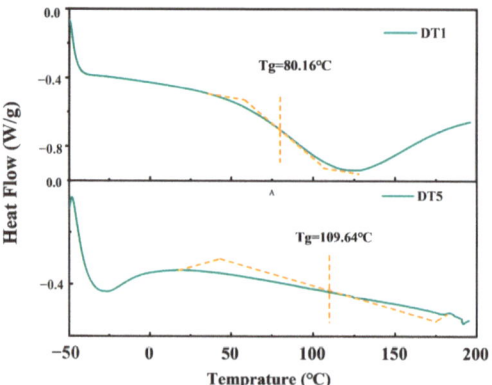

Figure 2. DSC curves of the DTMS/TEOS hybrid materials.

2.1.3. Micromorphology and Porosity

The specific surface and porosity of the optimally formulated hybrid material were analyzed. The specific surface area of the DT5 agent was found to be -0.0699 m^2/g. When combined with the N$_2$ adsorption–desorption graphs (Figure 3b), we observed that within the tested range, the hybrid material gel did not contain micropores or mesopores.

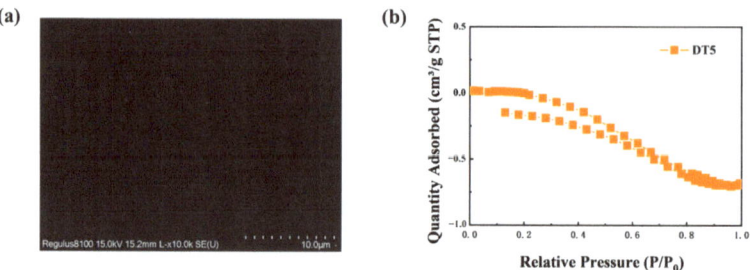

Figure 3. (a) SEM image under 10,000× magnification and (b) N2 adsorption–desorption curve of the TEOS/DTMS hybrid material.

The microstructure of the DT5 gel was observed under a scanning electron microscope. As revealed in Figure 3a, the surface of the polymer exhibited a flat, wrinkle-free, crack-free, dense, and uniform plate-like structure under 10,000× magnification. These characteristics indicated that the hybrid material lacked corresponding pores, which aligned with the porosity test results. We observed that DTMS mitigated the propensity of TEOS-based gel to cracking, and this was achieved, in part, by introducing flexible chains that could withstand capillary stress during the shrinkage process.

This dense and non-porous hybrid material gel, if uniformly encapsulated on the paper surface or grown in the pores formed by the fibers, could block the invasion of most external pollutants into the paper, such as dust, vermin, and mold [32,33].

2.2. Effectiveness of DTMS/TEOS Hybrid Materials for the Reinforcement of Aged Bamboo Paper

2.2.1. Appearance and Surface Hydrophobicity

As shown in Figure 4, the color change in DT5 at the three concentrations (10%, 30%, and 50%) tested on aged bamboo paper was generally indistinct, according to the color difference assessment level [34]. In particular, at a low concentration of 10%, DT5 hardly alters the color of the paper. We observed that as the concentration of the agent increased, more produced polymer coated the fiber surface and filled the pores, leading to a noticeable color difference in the paper. When the concentration of DT5 reached 50%, the paper demonstrated a distinctive, transparent texture. This was because the initial roughness of the paper surface caused light to scatter, resulting in a very low visual gloss on the surface. However, with the deposition of 50% DT5, the depressions between the fibers were filled. As a result, the smoothness of the paper surface and its gloss were significantly enhanced. Therefore, even though a high concentration of DT5 may better fill the pores and cracks on the paper surface, it is not recommended to exceed a usage concentration of 30% in order to avoid potential significant detrimental effects on the appearance of the paper relics.

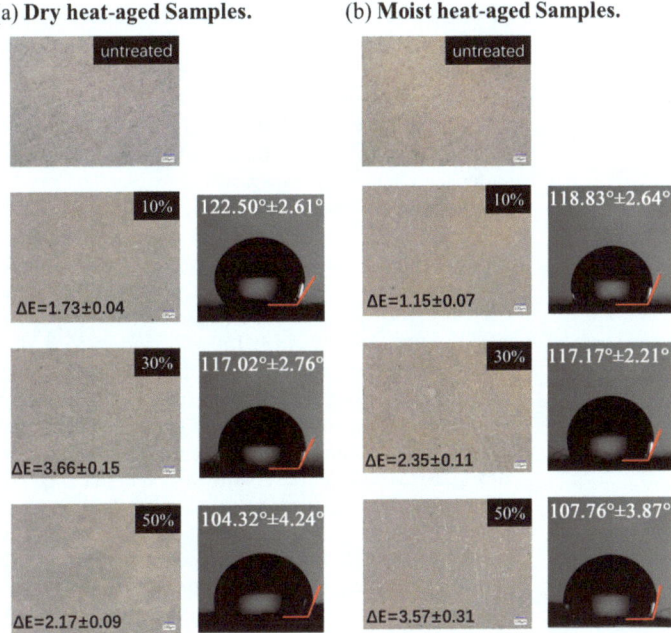

Figure 4. Changes in color and surface properties of the aged paper samples before and after treatment with DT5 agents in different concentrations. (**a**) Dry heat-aged samples; (**b**) moist heat-aged samples.

Cellulose, rich in hydroxyl groups, is highly hydrophilic and prone to absorbing moisture. Therefore, paper may over-absorb water in high environmental humidity conditions, leading to cellulose deliquescence and a decrease in the strength of its structure and organization. In addition, high humidity can facilitate the atmospheric deposition of hazardous dust and gases soluble in water, potentially corroding the paper. Furthermore, high humidity can promote the growth and reproduction of microorganisms, making it easy for mold to grow on paper [35]. To mitigate issues such as water stains, filth, adhesion, and wrinkles, it was essential that the protected paper exhibit hydrophobic properties that could block environmental water erosion.

Contact angle can serve as an indicator of paper surface hydrophobicity after treatment with a protective agent. When the contact angle decreased, the wettability of the solid surface improved and became more hydrophilic. Conversely, a larger contact angle indicated a more hydrophobic surface, signifying superior waterproof performance. We captured images of the contact angle on the surface of the paper after 10 s of exposure to water. Both the untreated sound bamboo paper and the two groups of aged paper lacked hydrophobicity, with water droplets being absorbed almost instantly; hence, no contact angle photos were obtained. In contrast, the treated aged bamboo paper consistently maintained pronounced hydrophobicity. As shown in Figure 4, the three concentrations of DT5 tested on aged bamboo paper had the ability to transform a fully wetted surface (contact angle of θ = 0°) into a non-wetted surface (contact angle of θ > 90°). This change highlighted the effectiveness of DTMS/TEOS hybrid materials in imparting hydrophobicity to aged bamboo paper. In summary, a 10% concentration of DT5 was sufficient to provide a hydrophobic surface for aged bamboo paper without altering its appearance.

2.2.2. Tensile Strength

The resistance of paper to aging will significantly impact its service life, making it crucial to investigate its aging resistance, with mechanical properties serving as key evaluation indicators. The most commonly used metric for assessing the mechanical properties of paper is tensile strength [36–38]. Therefore, to enhance the aging resistance of paper, the tensile strength of paper treated with a protective agent should be increased to a certain extent.

The mean and standard deviation of the tensile strength of aged bamboo paper before and after DT5 agent treatment are presented in Figure 5a, while the rate of change in tensile strength is shown in Figure 5b. The dry heat-aged bamboo paper exhibited longitudinal and transverse tensile strengths of 21.13 ± 0.31 MPa and 17.47 ± 0.39 MPa, respectively. Conversely, the moist heat-aged bamboo paper showed longitudinal and transverse tensile strengths of 25.99 ± 0.79 MPa and 13.26 ± 0.19 MPa, respectively. Notably, when the concentration of DT5 was 10%, the most significant enhancement effect was observed. In the dry heat-aged bamboo paper, the longitudinal and transverse tensile strengths were 25.48 ± 0.50 MPa and 24.87 ± 0.75 MPa, increasing by 20.59% and 36.63%, respectively. Similarly, in the moist heat-aged bamboo paper, the tensile strength values were 37.49 ± 0.35 MPa and 18.10 ± 0.76 MPa, increasing by 44.25% and 36.50%, respectively. The untreated sound paper demonstrated a longitudinal tensile strength of 27.54 ± 0.46 MPa and a transverse tensile strength of 21.03 ± 0.29 MPa. It is evident that 10% DT5 was sufficient to restore the strength of aged bamboo paper to a level comparable to or even higher than that of sound bamboo paper. However, when the concentration of DT5 increased, the increase in tensile strength values for both types of aged bamboo paper decreased in both directions. At a DT5 concentration of 50%, the tensile strength even dropped below the initial value. This finding aligns with the previous conclusion that 10% DT5 exhibited the most favorable enhancement effect, while higher concentrations of DT5 were counterproductive for the preservation of paper artifacts.

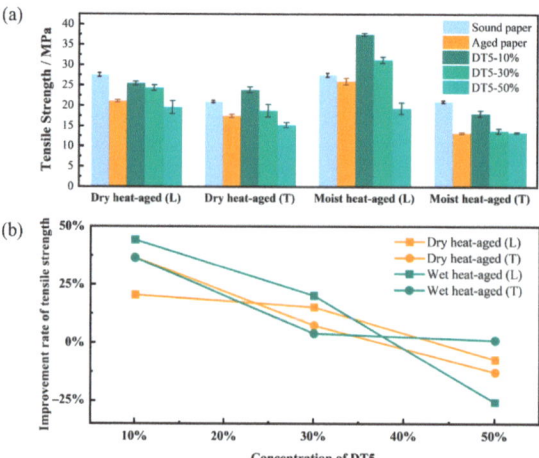

Figure 5. (a) Mean value and standard deviation of the tensile strength, and (b) change rate of the tensile strength of aged bamboo paper before and after treatment with DT5 agents at different concentrations, where T denotes the transverse direction and L signifies the longitudinal direction.

2.2.3. Microstructure

Scanning electron microscopy (SEM) analysis of the paper samples was conducted to observe the distribution of the hybrid materials, as illustrated in Figure 6. Bamboo paper primarily consists of slender fibers intertwined, along with a few flat and wide fibers. Prior to the protection treatment, the bamboo paper fibers exhibited cracks, and the interlacing connections between fibers were loose, with some of the interlacing connected by membranes, presumably due to the viscous additives used in the papermaking process [15]. After treatment with DT5 agents, the fibers were uniformly coated with the protective agent, resulting in a smoother surface. Block-like and membrane-like polymers formed at the intersections of the fibers, resulting in a more intimate connection between the fibers. This observation at the microscale might explain the reason behind the enhanced tensile strength of aged bamboo paper when treated with 10% and 30% DT5.

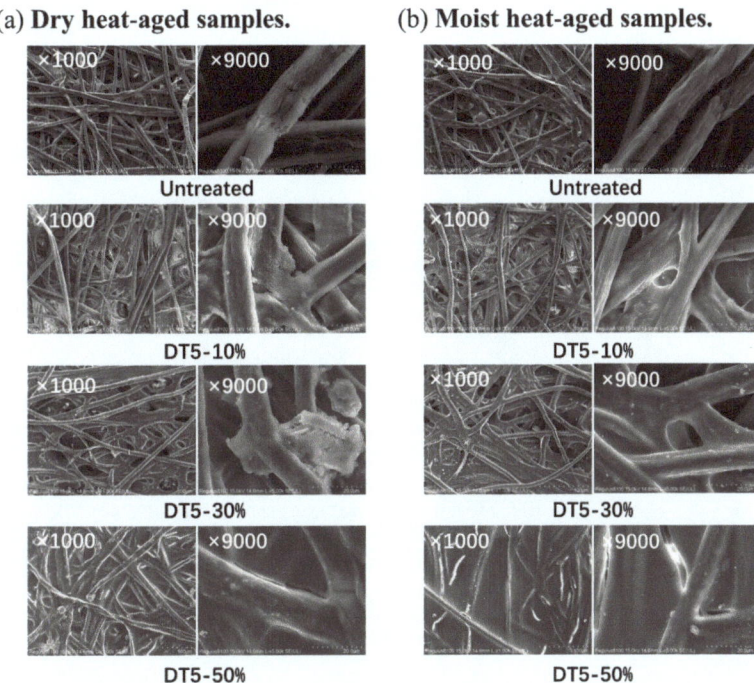

Figure 6. Microstructure of the aged bamboo paper before and after treatment with DT5 agents at different concentrations. (**a**) Dry heat-aged samples; (**b**) moist heat-aged samples.

However, when the concentration of DT5 reached 50%, the fibers were almost completely covered by the hybrid material, and the fiber pores were basically non-existent. We speculated that a high concentration of agents would lead to the formation of excessive hybrid material polymers inside the paper or even on the surface of the paper. The produced silicone polymer became the dominant component of the paper after curing; however, it did not have the interwoven fiber microstructures of sound paper, which provided high flexibility. Furthermore, when the agent concentration was excessively high, it could not only distribute within the gaps but also extensively infiltrate into the fibers, resulting in fiber damage during the curing process. As a result, the excessive polymers made paper into a resin-like film and changed the appearance as well as the mechanical properties of the paper.

2.2.4. Water Vapor Transmission Rate

The mass–time curves of the aged paper before and after treatment with 10% DT5 agent, obtained from the air permeability test, are shown in Figure 7. At relative humidity (RH) values below 25%, the curve was linear, allowing for the accurate computation of the water vapor transmission rate (WVTR). When the RH increased to 55%, the linearity of the curve persisted in the initial half, allowing for WVTR calculations. However, the second half of the curve started to approach a state of equilibrium, indicating that the molecular sieve reached saturation and ceased to absorb water. Similarly, at an RH value of 85%, the equilibrium portion of the curve could not be utilized for WVTR calculations due to the same saturation phenomenon.

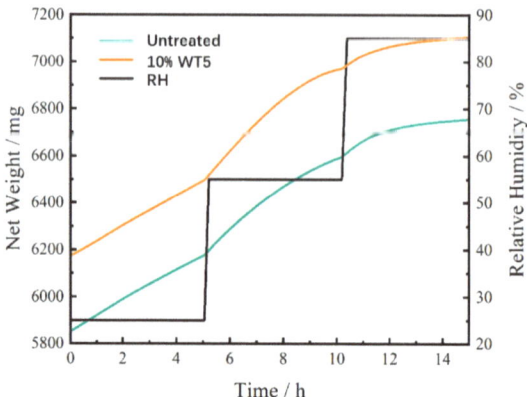

Figure 7. Mass–time curve of the paper sample before and after treatment with 10% DT5 agent.

At 25% RH, the linear mass change between 1.9 and 3.7 h was selected to calculate the WVTR, and at 55% RH, the corresponding interval between 5.4 and 6.1 h was chosen. The results of these calculations are summarized in Table 2, which indicates that regardless of treatment, the aged paper samples exhibited a high water vapor transmission rate. Notably, the DT5 agent had no significant impact on the air permeability of the paper.

Table 2. WVTR values of 10% DT5-treated aged paper at 25% and 55% RH determined by air permeability testing.

RH	Group	W_{start} (mg)	W_{end} (mg)	T (h)	WVTR [g/(m²·d)]
25%	Untreated	5982.087	6098.696	1.8	443.718
	10%DT5	6296.096	6413.247	1.8	445.780
55%	Untreated	6217.031	6292.713	0.7	740.528
	10%DT5	6541.522	6627.535	0.7	841.614

For immovable paper artifacts such as historical decorative wall coverings and paper-based murals, they are significantly influenced by moisture migration from the walls to which they were adhered. As the walls absorb water from the ground, they may create a highly humid environment on the backside of the attached paper and cause it to become damp. A high WVTR facilitates the release of moisture from the wall through the paper into the open environment rather than allowing it to accumulate or be absorbed into the fibers, leading to mold formation. Hence, while 10% DT5 did partially fill the pore structure of the paper, it did not transform it into a completely sealed polymer film. The treated paper still allows moisture to pass through, thereby preventing excessive moisture-related risks.

3. Materials and Methods

3.1. Preparation of the Hybrid Materials

The organosilanes employed in this study consisted of tetraethoxysilane (Aladdin, Shanghai, China, 99.99%) and dodecyltrimethoxysilane (Aladdin, Shanghai, China, 93%). DBTL (Aladdin, Shanghai, China, 95%) was used as a catalyzer, and absolute ethanol (Aladdin, Shanghai, China, AR) was used as the solvent.

The DTMS/TEOS hybrid material agents were prepared according to the formulations provided in Table 3 and thoroughly mixed to obtain a homogeneous mixture. Then, 2 mL of the prepared DTMS/TEOS sol with 0.2 wt% of DBTL was added to a plastic centrifuge tube, weighed, and placed inside the temperature and humidity environmental test chamber (GSH-64, ESPEC, Osaka, Japan). The conditions were set as follows: 25 °C and 55% RH for 3 h, followed by 25 °C and 85% RH for another 3 h. This cycle was repeated every 6 h. The moisture curing process was carried out under these conditions until the mass of the material stopped changing, indicating that the reaction had finished. The resulting gel was removed from the centrifuge tube and observed as transparent and free of cracks; then, its mass retention and volume shrinkage were calculated. The entire preparation process is illustrated in Figure 8.

Table 3. Formulations of the DTMS/TEOS hybrid material agents.

Group	W_{DTMS}/W_{TEOS}
DT1	1:9
DT2	2:8
DT3	3:7
DT4	4:6
DT5	5:5

Figure 8. Diagram of DTMS/TEOS hybrid material agent preparation.

3.2. Paper Aging

The handmade bamboo paper for artificial aging, which is widely used in calligraphy and painting creation, restoration of ancient books, and other paper relics, was produced by Fuyang Yiguzhai Yuanshu Paper Co., Ltd. (Fuyang, Zhejiang, China). The basis weight of the paper is 19.0 g·cm^{-2}, and the thickness is 0.03 mm. The bamboo paper was artificially aged by two aging methods, namely, dry heat aging and moist heat aging. The dry heat aging method was conducted according to the Chinese GB/T 464-2008 national standard [39]. The bamboo paper samples were subjected to dry heat aging in an electric constant-temperature blast-drying oven (DHG-9053A, Jinghong, China) under the following dry heat aging conditions: 105 °C and 24 days.

The moist heat aging method was conducted according to GB/T 22894-2008 [40]. The bamboo paper samples were subjected to moist heat aging in an environmental test chamber (SH-222, ESPEC, Osaka, Japan) under the following moist heat aging conditions: 80 °C, 65% RH, and 24 days.

3.3. Reinforcement of Paper Samples

The prepared protective solution was diluted with ethanol at various concentrations ($V_{solution}:V_{ethanol}$ = 1:9, 3:7, and 5:5) and then thoroughly mixed. The resulting agent was then evenly sprayed onto the aged bamboo paper using spray bottles. To process 0.1 m² of paper, approximately 1 milliliter of solution was used. Afterward, the treated paper was placed in an environmental test chamber (SH-222, ESPEC, Osaka, Japan) for moisture curing. The curing conditions were as follows: 25 °C and 55% RH for 3 h, followed by 25 °C and 85% RH for another 3 h. This cycle was then repeated every 6 h. The paper was left to dry for 1–2 days, and the moisture curing process was considered complete. The treating process is illustrated in Figure 9.

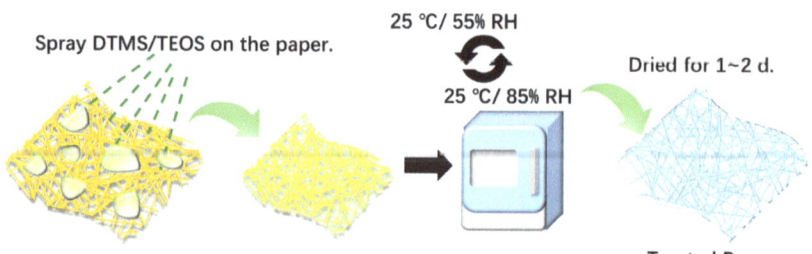

Figure 9. Diagram of the aged paper reinforcement process using DTMS/TEOS hybrid materials.

3.4. Specific Surface Area and Porosity

The pore structure of the materials was characterized by N_2 adsorption–desorption analysis. In this experiment, N_2 adsorption–desorption curves of DTMS/TEOS gels were obtained at 77 K using a specific surface and porosity analyzer (ASAP2460, Micromeritics, Norcross, Georgia, USA). The specific surface area of the samples was calculated based on the Brunauer–Emmet–Teller (BET) model, while the pore volume and average pore size of the samples were determined by analyzing the desorption data using the Barret–Joyner–Halenda (BJH) method.

3.5. Glass Transition Temperature

The glass transition temperature (Tg) can be defined as the point at which a material transitions from a glassy state to a highly elastic state. Below this temperature, the material will remain in a glassy state, with its internal molecular chains and chain segment movements effectively frozen. Once the temperature surpasses Tg, the material will enter a highly elastic state [31]. In practical applications, Tg can be used to determine the suitable temperature range for using the material. By assessing the Tg of the DTMS/TEOS gel, the thermal stability of the material within the desired use environment could be determined. For this experiment, a differential scanning calorimeter (DSC25, TA Instruments, New Castle, DE, USA) was used to analyze the Tg of the DTMS/TEOS hybrid materials. This method relied on detecting changes in the heat capacity before and after the glass transition to determine the magnitude of Tg, following the guidelines outlined in the GB/T 19466.2-2004 Plastics Differential Scanning Calorimetry (DSC) standard (Part 2: Determination of Glass Transition Temperature) [41].

3.6. Color Difference

The color difference (ΔE^*) between the aged paper samples before and after treatment with the DT5 agents was measured by a spectrophotometer (CM-26D, Konica Minolta, Japan) under a D65 light source. For each sample, the color difference was determined based on five points. The following equation was used to calculate the color difference ΔE^* for each point, and the average ΔE^* value was calculated to represent the ΔE^* of the entire sample [34]:

$$\Delta E^* = [(\Delta L^*)^2 + (\Delta a^*)^2 + (\Delta b^*)^2]^{1/2}, \tag{1}$$

where L^* ranges from 0 to 100 and indicates the change in color from black (dark) to white (light), a^* varies from negative to positive and represents the change in color from green to red, and b^* ranges from negative to positive and indicates the change in color from blue to yellow.

3.7. Contact Angle

In this experiment, a contact angle meter (JC2000DM, Zhongchen, China) was used to determine the contact angle and assess the hydrophobicity of the protected paper surface. Deionized water droplets were applied to the treated paper samples, and a fixed image was captured after 10 s. Three measurement points were taken for each sample, with a five-point fitting method used to measure the contact angle value at each point. The average of the three points was then calculated to determine the contact angle value of the sample.

3.8. Tensile Test

Tensile tests were conducted on the paper samples to assess their mechanical properties, and a thermomechanical analyzer (TMA7100, HITACHI, Tokyo, Japan) was utilized in this study. The TMA method enabled the acquisition of highly reproducible tensile strength data from millimeter-sized samples, making it particularly suitable for micro-destructive mechanical testing of fragile organic cultural relics [42]. Further details regarding this method can be found in our previous research paper [43]. The paper samples were cut into rectangular specimens measuring 15 mm × 2 mm in both the longitudinal and cross directions using a paper cutter. The thermomechanical analyzer (TMA) was equipped with a metal tensile attachment to conduct the tensile tests, where the test length, or fixture spacing (L), was set to 10 mm. The specimens were held at both ends using the fixtures, with the sample tubes and tensile probes firmly fixed in place. The tests were conducted in a laboratory room environment (25 ± 1 °C, 45% ± 3% RH) after the paper sample was conditioned in this test environment for more than 48 h with an initial load of 10 mN and a loading rate of 250 mN/min. The probe displacement and load were recorded over time until the specimen broke, and the stress (σ_f) and strain (ε_f) were calculated using Equations (2) and (3), respectively, to generate the stress–strain curve of the specimen:

$$\sigma_f = \frac{F}{bd}, \tag{2}$$

$$\varepsilon_f = \frac{\Delta L}{L}, \tag{3}$$

where σ_f is the fracture stress (MPa), F is the real-time load recorded by TMA (N), b is the width of the fracture surface of the specimen (mm), d is the thickness of the fracture surface of the specimen (mm), ε_f is the fracture strain, ΔL is the real-time displacement recorded by TMA (mm), and L is the specimen test length (mm), i.e., the fixture spacing.

The maximum stress at the time of fracture was recorded as the tensile strength σ, b was measured using a three-dimensional video microscope (VHX-6000, KEYENCE, Osaka, Japan) with an accuracy of 0.01 mm, and d was measured using a digital micrometer with an accuracy of 0.001 mm.

3.9. Microstructure

A scanning electron microscope (SEM) was used to examine the microstructure of the paper and DTMS/TEOS hybrid material gels. This analysis was used to assess fiber breakage in the paper, the pore structure and cracking of the hybrid material gels, and the distribution of the hybrid material polymers within the paper fibers. In this experiment, an ultra-high-resolution field-emission scanning electron microscope (Regulus 8100, HITACHI, Tokyo, Japan) was used. The surface of the paper samples was subjected to a gold spraying treatment. The microscope was operated with an accelerating voltage of 15 kV, a resolution of 0.7 nm, a working distance of 15 mm, and a secondary electron imaging (SE) working mode.

3.10. Water Vapor Transmission Test

A high-throughput dynamic moisture-adsorption tester (SPSx-1μ, ProUmid, Ulm, Germany) was utilized to assess the permeability of the samples under identical temperature conditions and various relative humidity levels. In the test setup, approximately 5 g of molecular sieve was placed in the sample tray. The bamboo paper was cut into a circular shape (radius = 4 cm) and securely attached to the fixture on the sample tray. The test parameters were set as follows: test range: 25–85% RH; humidity gradient: 30% RH; temperature: 25 °C. The change in the net weight of the sample over a specific period was determined, and the water vapor transmission rate (WVTR) of the bamboo paper was calculated using the following equation:

$$WVTR = \frac{W_{end} - W_{start}}{\pi R^2 \cdot T} \quad (4)$$

where W_{start} is the net weight of the paper sample at the selected start time (mg), W_{end} is the net weight at the selected end time (mg), R is the radius of the bamboo paper sample (which was 4 cm in this study), and T is the total time from the selected start time to the end time (h).

4. Conclusions

This study demonstrated that the DTMS/TEOS hybrid materials possessed excellent anti-cracking properties. The hybrid materials provided transparent and dense gels with no severe cracks. We observed that with the increase in DTMS addition, the resulting gels were less prone to cracking. In addition, these gels exhibited good thermal stability, which was attributed to the flexible chains introduced by DTMS, and this helped to resist capillary stress during shrinkage. The experimental results indicated that under the catalytic conditions of a 0.2% mass concentration of DBTL, the DTMS/TEOS hybrid materials achieved good curing effects across various mass fraction ratios. The DT5 agent (with a W_{DTMS}/W_{TEOS} = 5:5) achieved the highest mass retention rate of 63.23%. At this ratio, the volume shrinkage was minimal at 36.72%, and the complete curing time was 307 h, making it the most preferred formulation.

In addition, we observed that a 10–50% concentration of DT5 agent in the ethanol solution effectively reinforced and protected aged paper. Notably, there was no significant change in the color appearance of the treated paper, and its surface exhibited strong hydrophobicity. Among the various tested concentrations, the 10% concentration exhibited the best reinforcement effect. After dry heat aging, the tensile strength of the bamboo paper increased by 20.59% and 36.63% in the longitudinal and cross directions, respectively. Similarly, the moist heat-aged paper samples resulted in 44.25% and 36.50% increases in tensile strength in the longitudinal and cross directions, respectively. The DTMS/TEOS hybrid material formed a strong and dense network structure on the surface and between the fibers of the paper, filling the pores and imparting strong hydrophobicity to the paper. This pore closure significantly enhanced the mechanical properties and hydrophobicity of the paper while maintaining its air permeability almost entirely.

In conclusion, the developed DTMS/TEOS hybrid material demonstrated a positive reinforcement and preservation effect on paper cultural relics. In addition, the material demonstrated ease of preparation, low cost, and wide accessibility, making it a promising choice for future research and applications for paper relic preservation.

Author Contributions: Conceptualization, X.H. and H.G.; methodology, X.H.; software, M.W.; validation, X.H. and H.G.; formal analysis, L.M.; investigation, X.H., M.W. and Z.Z.; resources, L.M., H.G. and L.H.; data curation, X.H. and H.G.; writing—original draft preparation, M.W. and Z.Z.; writing—review and editing, X.H. and L.H.; visualization, M.W., L.M. and Z.Z.; supervision, X.H.

and H.G.; project administration, X.H., H.G. and L.H.; funding acquisition, H.G. and L.H. All authors have read and agreed to the published version of the manuscript.

Funding: This research was funded by the National Key R&D Program of China, grant numbers 2022YFF0903905 and 2022YFF0903902.

Institutional Review Board Statement: Not applicable.

Informed Consent Statement: Not applicable.

Data Availability Statement: The data presented in this study are available on request from the corresponding author.

Acknowledgments: The authors would like to thank Lifang Ji from the Palace Museum for providing the paper samples.

Conflicts of Interest: The authors declare no conflicts of interest.

References

1. Ipert, S.; Dupont, A.L.; Lavédrine, B.; Bégin, P.; Rousset, E.; Cheradame, H. Mass deacidification of papers and books. IV—A study of papers treated with aminoalkylalkoxysilanes and their resistance to ageing. *Polym. Degrad. Stab.* **2023**, *91*, 3448–3455. [CrossRef]
2. Xu, Y.; Yin, H.; Yuan, S.; Chen, Z. Film morphology and orientation of amino silicone adsorbed onto cellulose substrate. *Appl. Surf. Sci.* **2009**, *255*, 8435–8442. [CrossRef]
3. He, W.; Zhang, Z.; Zheng, Y.; Qiao, S.; Xie, Y.; Sun, Y.; Qiao, K.; Feng, Z.; Wang, X.; Wang, J. Preparation of aminoalkyl-grafted bacterial cellulose membranes with improved antimicrobial properties for biomedical applications. *J. Biomed. Mater. Res. Part A* **2020**, *108*, 1086–1098. [CrossRef] [PubMed]
4. Xie, Y.; Hill, C.A.S.; Xiao, Z.; Holger, M.; Carsten, M. Silane coupling agents used for natural fiber/polymer composites: A review. *Compos. Part A Appl. Sci. Manuf.* **2010**, *41*, 806–819. [CrossRef]
5. Gentle, T.E.; Schmidt, R.G.; Naasz, B.M.; Gellman, A.J.; Gentle, T.M. Organofunctional silanes as adhesion promoters: Direct characterization of the polymer/silane interphase. *J. Adhes. Sci. Technol.* **1992**, *6*, 307–316. [CrossRef]
6. Peter, G.P.; Edwin, P.P. Methods for improving the performance of silane coupling agents. *J. Adhes. Sci. Technol.* **1991**, *5*, 831–842. [CrossRef]
7. Yilgör, E.; Yilgör, I. Silicone containing copolymers: Synthesis, properties and applications. *Prog. Polym. Sci.* **2014**, *39*, 1165–1195. [CrossRef]
8. Abdelmouleh, M.; Boufi, S.; Belgacem, M.N.; Duarte, A.P.; Salah, A.B.; Gandini, A. Modification of cellulosic fibres with functionalised silanes: Development of surface properties. *Int. J. Adhes.* **2004**, *24*, 43–54. [CrossRef]
9. Oh, M.; Lee, S.; Paik, K. Preparation of hydrophobic self-assembled monolayers on paper surface with silanes. *J. Ind. Eng. Chem.* **2011**, *17*, 149–153. [CrossRef]
10. Nowak, T.; Mazela, B.; Olejnik, K.; Peplińska, B.; Perdoch, W. Starch-Silane Structure and Its Influence on the Hydrophobic Properties of Paper. *Molecules* **2022**, *27*, 3136. [CrossRef]
11. Koga, H.; Kitaoka, T.; Isogai, A. Chemically-Modified Cellulose Paper as a Microstructured Catalytic Reactor. *Molecules* **2015**, *20*, 1495–1508. [CrossRef] [PubMed]
12. Carla, G.; Ana, P.C.; Mario, N.; Manuel, J.S.S.; Mohamed, N.B. Grafting of paper by silane coupling agents using cold-plasma discharges. *Plasma Process. Polym.* **2008**, *5*, 444–452. [CrossRef]
13. Mahin, H.H.; Omid, R.; Habib, B. Silane–based modified papers and their extractive phase roles in a microfluidic platform. *Anal. Chim. Acta* **2020**, *1128*, 31–41. [CrossRef]
14. Chen, K.; Yang, Y.; Li, P.; Zhan, Y. Study on the Strengthening of Paper Relics by Aminoalkylalkoxysilane. *Guangdong Chem. Ind.* **2017**, *44*, 11–13+22. (In Chinese)
15. Shen, Y.; Li, Z.; Hou, A.; Chen, K.; Zhan, Y. Study on Reinforcement and Deacidification of Xuan Paper by Aminoalkylalkoxysilane Coupling Agents. *China Pulp Pap.* **2018**, *37*, 45–48. (In Chinese)
16. Dupont, A.L.; Lavédrine, B.; Cheradame, H. Mass deacidification and reinforcement of papers and books VI—Study of aminopropylmethyldiethoxysilane treated papers. *Polym. Degrad. Stab.* **2010**, *95*, 2300–2308. [CrossRef]
17. Souguir, Z.; Dupont, A.L.; d'Espinose de Lacaillerie, J.B.; Lavedrine, B.; Cheradame, H. Chemical and physicochemical investigation of an aminoalkylalkoxysilane as strengthening agent for cellulosic materials. *Biomacromolecules* **2011**, *12*, 2082–2091. [CrossRef]
18. Souguir, Z.; Dupont, A.L.; Fatyeyeva, K.; Mortha, G.; Cheradame, H.; Ipert, S.; Lavédrine, B. Strengthening of degraded cellulosic material using a diamine alkylalkoxysilane. *RSC Adv.* **2012**, *2*, 7470–7478. [CrossRef]
19. Piovesan, C.; Dupont, A.L.; Fabre-Francke, I.; Fichet, O.; Lavédrine, B.; Chéradame, H. Paper strengthening by polyaminoalkylalkoxysilane copolymer networks applied by spray or immersion: A model study. *Cellulose* **2014**, *21*, 705–715. [CrossRef]
20. Piovesan, C.; Fabre-Francke, I.; Paris-Lacombe, S.; Dupont, A.L.; Fichet, O. Strengthening naturally and artificially aged paper using polyaminoalkylalkoxysilane copolymer networks. *Cellulose* **2018**, *25*, 6071–6082. [CrossRef]

21. Natali, I.; Tomasin, P.; Becherini, F.; Bernardi, A.; Ciantelli, C.; Favaro, M.; Favoni, O.; Pérez, V.J.F.; Olteanu, I.D.; Sanchez, M.D.R.; et al. Innovative consolidating products for stone materials: Field exposure tests as a valid approach for assessing durability. *Herit. Sci.* **2015**, *3*, 6. [CrossRef]
22. Luo, H.; Liu, R.; Huang, X. Progress in Research of Crack-Free Organosilane Consolidants for Stone Conservation. *Mater. China* **2012**, *31*, 1–8. (In Chinese)
23. Maravelaki-Kalaitzaki, P.; Kallithrakas-Kontos, N.; Korakaki, D.; Agioutantis, Z.; Maurigiannakis, S. Evaluation of silicon-based strengthening agents on porous limestones. *Prog. Org. Coat.* **2006**, *57*, 140–148. [CrossRef]
24. Zárraga, R.; Cervantes, J.; Salazar-Hernandez, C.; Wheeler, G. Effect of the addition of hydroxyl-terminated polydimethylsiloxane to TEOS-based stone consolidants. *J. Cult. Herit.* **2010**, *11*, 138–144. [CrossRef]
25. Mosquera, M.J.; de los Santos, D.M.; Rivas, T. Surfactant-synthesized ormosils with application to stone restoration. *Langmuir* **2010**, *26*, 6737–6745. [CrossRef] [PubMed]
26. Luo, Y.; Xiao, L.; Zhang, X. Characterization of TEOS/PDMS/HA nanocomposites for application as consolidant/hydrophobic products on sandstones. *J. Cult. Herit.* **2015**, *16*, 470–478. [CrossRef]
27. Miliani, C.; Velo-Simpson, M.L.; Scherer, G.W. Particle-modified consolidants: A study on the effect of particles on sol-gel properties and consolidation effectiveness. *J. Cult. Herit.* **2007**, *8*, 1–6. [CrossRef]
28. Liu, R.; Han, X.; Huang, X.; Li, W.; Luo, H. Preparation of three-component TEOS-based composites for stone conservation by sol-gel process. *J. Sol-Gel Sci. Technol.* **2013**, *68*, 19–30. [CrossRef]
29. Xu, F.; Zeng, W.; Li, D. Recent advance in alkoxysilane-based consolidants for stone. *Prog. Org. Coat.* **2019**, *127*, 45–54. [CrossRef]
30. Son, S.; Won, J.; Kim, J.; Jang, Y.; Kang, Y.; Kim, S. Organic–Inorganic Hybrid Compounds Containing Polyhedral Oligomeric Silsesquioxane for Conservation of Stone Heritage. *ACS Appl. Mater. Interfaces* **2009**, *1*, 393–401. [CrossRef]
31. Negi, A.; Sumit, B. A molecular dynamics study on the strength and ductility of high Tg polymers. *Model. Simul. Mater. Sci. Eng.* **2006**, *14*, 563. [CrossRef]
32. Li, Q.; Xi, S.; Zhang, X. Conservation of paper relics by electrospun PVDF fiber membranes. *J. Cult. Herit.* **2014**, *15*, 359–364. [CrossRef]
33. Li, Q. Advanced Technology for Deacidification and Conservation of Paper Relics. Doctoral Dissertation, Zhejiang University, Hangzhou, China, 2014.
34. *GB/T 7921-2008*; Uniform Color Space and Color Difference Formula. Standardization Administration of the People's Republic of China, General Administration of Quality Supervision, Inspection and Quarantine of the People's Republic of China, Standards Press of China: Beijing, China, 2008.
35. Bogaard, J.; Whitmore, P. Explorations of the Role of Humidity Fluctuations in the Deterioration of Paper. *Stud. Conserv.* **2002**, *47*, 11–15. Available online: https://api.semanticscholar.org/CorpusID:95985118 (accessed on 1 September 2002). [CrossRef]
36. Chen, B.; Tan, J.; Huang, J.; Lu, Y.; Gu, P.; Han, J.; Ding, Y. Research on the aging-resistance properties of four kinds of Fuyang bamboo paper. *J. For. Eng.* **2021**, *6*, 121–126. (In Chinese)
37. Tian, Z.; Yan, Z.; Ren, S.; Yi, X.; Long, K.; Zhang, M. Research on the Dry Heat Aging Resistant Properties of Different Papers. *China Pulp Pap.* **2017**, *36*, 42–47. (In Chinese)
38. Zhang, N.; Zheng, D.; He, W.; Min, H. Study on the Properties of Restoration Paper Used in Suzhou Style Mounting. *China Pulp Pap.* **2019**, *38*, 79–84. (In Chinese)
39. *GB/T 464-2008*; Paper and Board—Accelerated Aging—Dry Heat Treatment. Standardization Administration of the People's Republic of China, General Administration of Quality Supervision, Inspection and Quarantine of the People's Republic of China, Standards Press of China: Beijing, China, 2008.
40. *GB/T 22894-2008*; Paper and Board—Accelerated Ageing—Moist Heat Treatment at 80 °C and 65% Relative Humidity. Standardization Administration of the People's Republic of China, General Administration of Quality Supervision, Inspection and Quarantine of the People's Republic of China, Standards Press of China: Beijing, China, 2008.
41. *GB/T 19466.2-2004*; Plastics—Differential Scanning Calorimetry (DSC)—Part 2: Determination of Glass Transition Temperature. Standardization Administration of the People's Republic of China, General Administration of Quality Supervision, Inspection and Quarantine of the People's Republic of China. Standards Press of China: Beijing, China, 2004.
42. Wu, M.; Han, X.; Qin, Z.; Zhang, Z.; Xi, G.; Han, L. A Quasi-Nondestructive Evaluation Method for Physical-Mechanical Properties of Fragile Archaeological Wood with TMA: A Case Study of an 800-Year-Old Shipwreck. *Forests* **2022**, *13*, 38. [CrossRef]
43. Zhang, Z.; Zhang, W.; Han, X. Evaluation of the Aging Property of Bamboo Paper Used for the Restoration of Pengbihushi in the Palace Museum. *Spectrosc. Spect. Anal.* **2023**, *43*, 1968–1973. Available online: https://www.gpxygpfx.com/EN/10.3964/j.issn.1000-0593(2023)06-1968-06 (accessed on 20 June 2023).

Disclaimer/Publisher's Note: The statements, opinions and data contained in all publications are solely those of the individual author(s) and contributor(s) and not of MDPI and/or the editor(s). MDPI and/or the editor(s) disclaim responsibility for any injury to people or property resulting from any ideas, methods, instructions or products referred to in the content.

Article

Facile Synthesis of Low-Dimensional and Mild-Alkaline Magnesium Carbonate Hydrate for Safe Multiple Protection of Paper Relics

Yi Wang [1,2], Zirui Zhu [1,3], Jinhua Wang [2], Peng Liu [1,3], Xingxiang Ji [1], Hongbin Zhang [1,2,3,*] and Yi Tang [4,*]

1. State Key Laboratory of Biobased Material and Green Papermaking, Qilu University of Technology, Shandong Academy of Sciences, Jinan 250353, China; ebanwang@163.com (Y.W.); 22110820002@m.fudan.edu.cn (Z.Z.); liupengfdu@fudan.edu.cn (P.L.); xxjt78@163.com (X.J.)
2. Department of Cultural Relics and Museology, Fudan University, Shanghai 200433, China; jinhuawang@fudan.edu.cn
3. Institute for Preservation and Conservation of Chinese Ancient Books, Fudan University Library, Fudan University, Shanghai 200433, China
4. Department of Chemistry, Laboratory of Advanced Materials, Collaborative Innovation Center of Chemistry for Energy Materials and Shanghai Key Laboratory of Molecular Catalysis and Innovative Materials, Fudan University, Shanghai 200433, China
* Correspondence: zhanghongbin@fudan.edu.cn (H.Z.); yitang@fudan.edu.cn (Y.T.)

Citation: Wang, Y.; Zhu, Z.; Wang, J.; Liu, P.; Ji, X.; Zhang, H.; Tang, Y. Facile Synthesis of Low-Dimensional and Mild-Alkaline Magnesium Carbonate Hydrate for Safe Multiple Protection of Paper Relics. *Molecules* **2024**, *29*, 4921. https://doi.org/10.3390/molecules29204921

Academic Editor: Maria Luisa Saladino

Received: 12 September 2024
Revised: 11 October 2024
Accepted: 16 October 2024
Published: 17 October 2024

Copyright: © 2024 by the authors. Licensee MDPI, Basel, Switzerland. This article is an open access article distributed under the terms and conditions of the Creative Commons Attribution (CC BY) license (https://creativecommons.org/licenses/by/4.0/).

Abstract: Paper-based cultural relics inevitably face a variety of diseases such as acidification, yellowing, and strength loss during long-term preservation, where weakly alkaline inorganic materials play an important role in their deacidification treatments. In this work, by simply adjusting the supersaturation of crystal growing solution without the use of any organic additives, one-dimensional (1D) and two-dimensional (2D) weakly alkaline materials—magnesium carbonate hydrates (MCHs)—were controllably synthesized. It is worth noting that the coatings of 1D/2D MCHs not only cause little change in chromatic aberration and water wettability, but also ensure their safety for alkali-sensitive pigments. Meanwhile, the deacidification, anti-aging, strength-enhancing, and flame-retardant effects of these materials have been tested on ancient book papers, all of which achieved good protective effects. In contrast, 1D MCH materials brought about significant enhancement in both mechanical strengths and flame-retardant effects, and the related effects were investigated. Based on this facile micromorphology control strategy, more low-dimensional nanomaterials are expected to be synthesized by design for the protection of paper-based relics, which will expand our knowledge on functional deacidification and protection mechanisms.

Keywords: low-dimensional materials; mild alkalinity; protection of paper relics; safe deacidification; paper enhancement

1. Introduction

As one of the most important carriers of cultural information, paper-based relics play a vital role in the history of mankind. There exists inestimable historical, artistic, and scientific value in paintings, calligraphy works, archives, books, and other kinds of paper-based cultural relics. Although much importance has been attached to these precious tangible cultural heritages, deterioration unavoidably occurs during their storage. Naturally, as the main component of paper, cellulose undergoes a chain degradation reaction inevitably, which leads to the acidification of paper, causing yellowing, discoloration, strength decline, and other problems that threaten the longevity of paper relics [1,2]. To avoid the adverse consequences of accumulating acidity, deacidification has been considered the most crucial step in conservation of paper relics. Studies show that hydroxides or oxides are mainly employed as deacidification agents (such as calcium hydroxide [3–5] and magnesium oxide [6–8]). However, the excessive alkalinity of deacidification materials and the iterative treatment process may bring some side effects, causing damage to the fibers and thus reducing the mechanical properties of papers, which obviously go against the safety requirements for fragile paper relic protection [9,10]. Therefore, the mild alkaline, efficient, and multifunctional protective material is still desirable.

In recent decades, low-dimensional materials and their composites have received great research attention in terms of cellulose-based material enhancement and multi-functionalization [11–13] due to their diverse properties, and are expected to play a special role in paper relic protection. One-dimensional (denoted as 1D) materials usually exhibit typical cross-sectional dimensions in the nanoscale and lengths spanning from hundreds of nanometers to millimeters. Among them, whiskers are a special class of one-dimensional materials, which are short fibers grown from single crystals with a highly ordered atomic arrangement and fine diameters, and whose strength is close to the theoretical value of a complete crystal [14]. As in the paper industry, magnesium salt whiskers have been used as a filler to enhance the flame-retardant performance of paper, and calcium carbonate whiskers were proved to be effective in improving the physical properties of paper [15,16]. Two-dimensional (denoted as 2D) nanomaterials are layered compounds with a nanometer-scale thickness, large specific surface area, and a high density of surface-active sites capable of superior flexibility, transparency, and breathability. Existing research suggests that two-dimensional zeolite materials can not only provide sufficient, mild alkaline sites, but also improve the mechanical strength of the paper through film formation [17]. Due to the superior specific surface area and tunable surface chemistry, 2D materials can provide continuous protection as a good carrier, which helps to enhance the aging resistance of paper-based artifacts [18–20]. Nevertheless, although these low-dimensional synthetic materials exhibit excellent properties in paper-based materials, their synthesis process is complex and time-consuming, and organic templates and capping agents are usually used, which may limit their widespread use in the protection of paper relics.

Perhaps the simplest and most cost-effective manner to prepare low-dimensional inorganic materials is through precipitation from a solution. This process involves a nucleation and a growth stage. As the nuclei formed approach a critical concentration, additional reactants can either attach to the existing nuclei, leading to particle growth, or form new nuclei [21,22]. Naturally, the morphology of the precipitates produced depends on which of these two competitive steps prevails, although they all proceed quickly, the growth of particles and materials with the desired morphology can be facilitated by finding reasonable control conditions. Such synthetic methods to control the product morphology by adjusting the degree of ionization and supersaturation have also been reported in the literature in recent years [23]. Thus, based on this type of method, we expect to see more development of weakly alkaline materials for the protection of paper-based artifacts.

Among the weakly basic inorganic compounds with tunable micromorphology, magnesium carbonate hydrates are a class of important materials, and the general formula can be expressed as $xMgCO_3 \cdot yMg(OH)_2 \cdot zH_2O$ (x = 1–5, y = 0–1, z = 1–8). To date, a number of synthesis methods have been developed to generate diverse morphologies [24–27]. However, the simple and low-cost controllable synthesis of low-dimensional magnesium carbonate hydrate materials still needs to be developed, and its new applications such as paper relic protection are not yet covered. In this paper, low-dimensional weakly alkaline magnesium carbonate hydrates with different morphologies (1D whisker and 2D nanosheet) were prepared by adopting a synthetic strategy of controlling crystal growth through solution supersaturation without using any organic additives, and the materials were tested for the coating effect and safety, as well as for their protective effects on historical wood-pulp papers (denoted as HMP). Both protective materials showed good deacidification and anti-aging properties, and their different structure and composition enabled them to play differentiated roles in the mechanical strength enhancement and flame-retardant performance improvement of paper. As shown in Scheme 1, this work provides a brand new perspective on the study of paper conservation materials and shows bright prospects in the protection of diverse cultural relics.

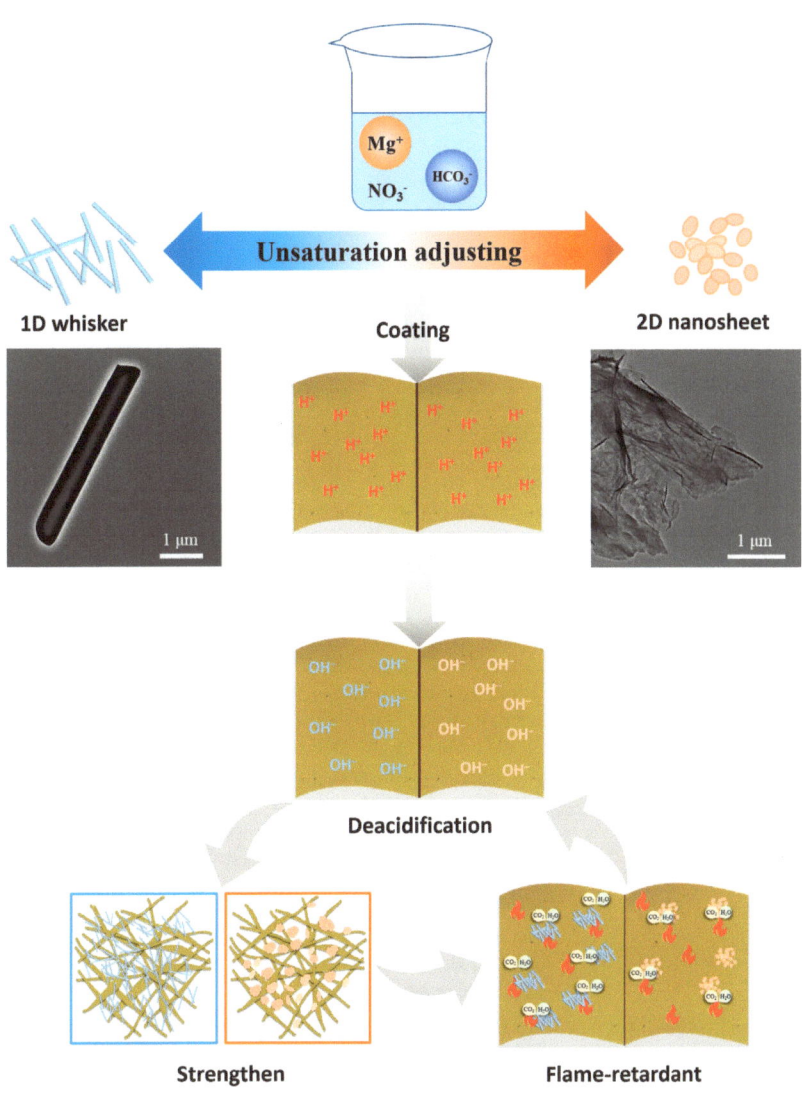

Scheme 1. Illustration of micromorphology modulation of materials and their protection effect on paper relics.

2. Results and Discussion

2.1. Structure and Properties of 1D and 2D Products

The SEM and TEM results (Figure 1) clearly demonstrate that the synthesis method employed in this work easily yields good 1D and 2D microstructure. The diameter of 1D whisker is mainly between 0.5 and 1 μm, and the length can be larger than 30 μm. It can be seen from Figure 1(b_2) that the thickness of 2D nanosheets is very thin, about 5 nm. The related element mapping data preliminarily indicate that the elemental composition of the two products is characteristic of the magnesium carbonate hydrate compounds. This result accords with the expectation of supersaturation adjustment design.

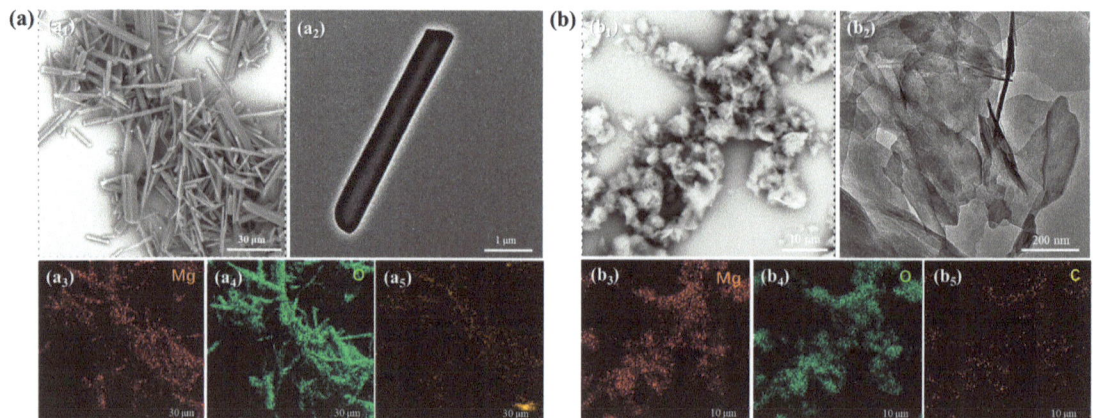

Figure 1. (a_1) SEM image, (a_2) TEM image, and (a_3–a_5) related element analysis of 1D product; (b_1) SEM image, (b_2) TEM image, and (b_3–b_5) related element analysis of 2D product.

The XRD pattern (Figure 2a) reveals that both products with different morphologies are typically magnesium carbonate hydrates, and the phases of 1D and 2D products are $MgCO_3 \cdot nH_2O$ and $4MgCO_3 \cdot Mg(OH)_2 \cdot nH_2O$, respectively [28,29], denoted as 1D-MCH and 2D-MCH. The peak style of the (101), (200), (400), (021), and (004) characteristic crystal faces of 1D-MCH are sharp, indicating the good crystallinity of the sample. However, the diffraction peaks corresponding to the characteristic crystal plane (100), (011), (−102), (310), and (−413) of 2D-MCH are relatively wide, influenced by the small size of the sample layer. Further, the FT-IR data are tested to study the state of characteristic groups inside these samples. The absorption peaks at 3200–3500 cm^{-1} in the IR spectrum (Figure 2b) are characteristic of water and -OH in the sample, while the stronger double peaks at 1476 and 1414 cm^{-1} are asymmetrical contraction vibrations of the carbonate, and the weaker absorption at 1117 cm^{-1} is the symmetrical stretching vibration of the carbonate. The similar characteristic regions in the IR spectra of the two samples indicate that they have a close composition.

In addition, a thermogravimetric analysis (TGA) is adopted to explore the thermal decomposition and compositions of these samples, as shown in Figure 2c,d. 1D-MCH is essentially stable in a mass up to 100 °C, with its first peak in the DSC curves at 180 °C due to the loss of crystal water, and the mass loss in this stage is 32.1%. In contrast, 2D-MCH is less thermally stable at low temperatures (<100 °C), with weight loss and endothermic in the early stage of heating, with 22.3% of mass loss in the first stage. The highest peak of DSC curves of 1D-MCH and 2D-MCH was observed at 391 and 387 °C, respectively, indicating the release of CO_2. Thus, in the second stage, the mass loss for 1D-MCH and 2D-MCH is 34.7% and 35.5%, respectively, which is jointly composed of contributions from the release of carbon dioxide and water. At about 400 °C, the mass percentage of the two samples gradually tends to be stable, and finally, they will completely decompose into MgO. Based on the final mass percentage, it can be calculated that the average number of water molecules bound in 1D-MCH and 2D-MCH crystals is 2.1 and 4.6, respectively.

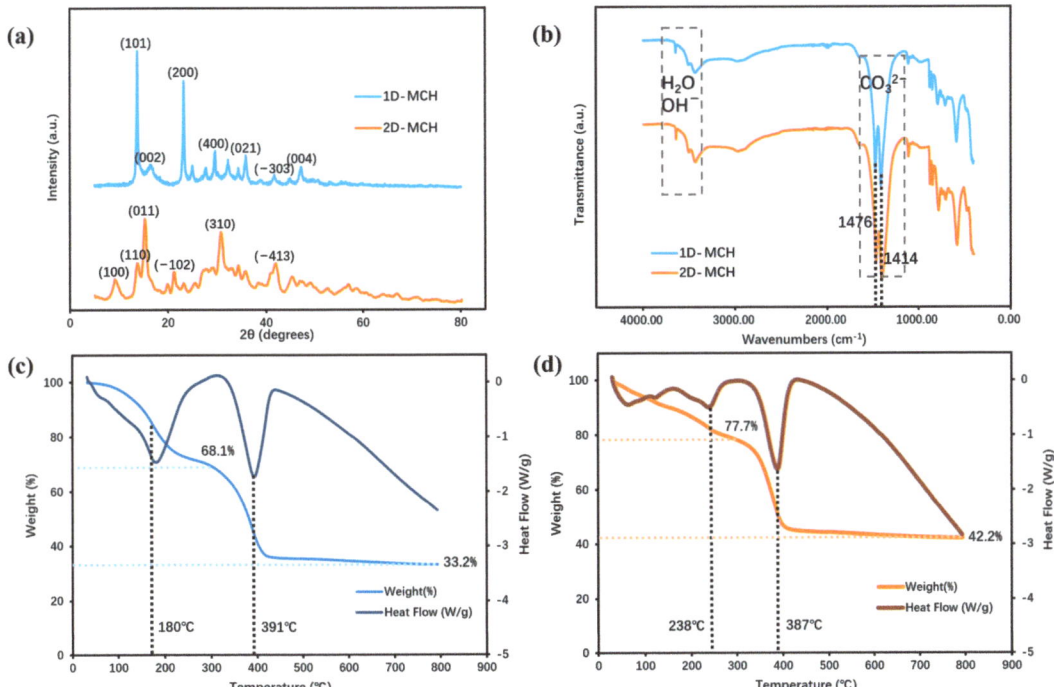

Figure 2. (a) XRD pattern and (b) FT-IR spectra of 1D and 2D products; (c) DSC-TGA curve of 1D product and (d) DSC-TGA curve of 2D product.

2.2. Coating Property and Safety Test

Before the application experiments, several tests were performed to evaluate the coating property and safety of protective materials, including the contact angle tests of the paper samples before and after coating and chroma changes during dry-heat aging. It can be seen from Figure 3 that the deposition of material formed a good microscopic filling between the paper fibers. Especially in the 1D-MCH-treated sample, the filling compactness between fibers was significantly improved. At the same time, the contact angle of paper samples had no significant change, which indicates that the treated paper still maintains a certain hydrophobicity. Furthermore, the results of the chromaticity change show that the effect of protection treatment on paper chromaticity is in an acceptable range (Table 1).

Table 1. CIE color coordinates of uncoated and coated HMP.

Sample (HMP)	L	a*	b*	ΔE*
uncoated	78.96	8.67	21.90	--
1D-MCH-coated	77.93	7.35	21.40	1.76
1D-MCH-coated, aged for 7 d	78.16	7.42	22.04	1.50
1D-MCH-coated, aged for 14 d	77.98	7.03	21.08	2.09
2D-MCH-coated	78.50	7.11	20.81	1.96
2D-MCH-coated, aged for 7 d	78.33	6.85	21.78	1.94
2D-MCH-coated, aged for 14 d	80.30	8.94	21.52	1.41

* These values are calculated based on the CIEDE2000 formulae.

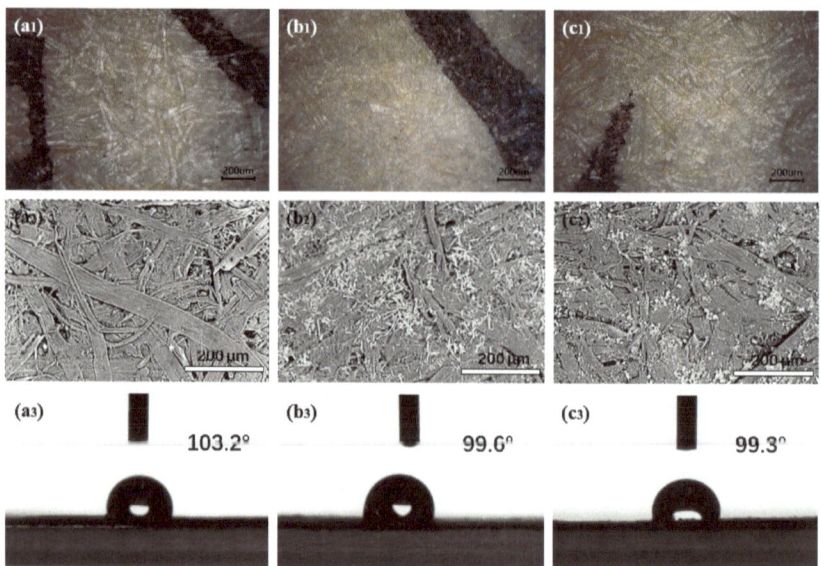

Figure 3. (a_1) Optical microscopy image, (a_2) SEM image and (a_3) contact angle of water droplets of uncoated HMP; (b_1) Optical microscopy image, (b_2) SEM image and (b_3) contact angle of water droplets of 1D-MCH-coated HMP; (c_1) Optical microscopy image, (c_2) SEM image and (c_3) contact angle of water droplets of 2D-MCH-coated HMP.

Among the various international conventions on the protection of cultural heritage, integrity and authenticity and are two main principles to be followed in the protection of tangible cultural heritage. When it comes to paper relics like paintings and calligraphy relics, one of the most important original states is the color of the pigments used. Therefore, it is an important indicator to evaluate the safety of alkaline deacidifiers used in paper deacidification protection that whether they will adversely affect the color of alkali-sensitive pigments. In this work, two pigments sensitive to alkali, lead chrome yellow ($PbCrO_4$) and Prussian blue ($Fe_4[Fe(CN)_6]_3$), were selected to test the influence of alkaline protection materials on their color.

As shown in Figure 4a, the color state of lead chrome yellow (LCY) changed slightly after accelerated aging. The study by Costana et al. proved that the reduction in Cr(VI) to Cr(III) caused by light should be the main reason for the aging and fading of lead chrome yellow in oil painting, which is also related to the purity of lead chromate in pigment [30]. However, as an important mineral pigment widely used in the Chinese art of painting during the Qing Dynasty (A.D. 1636–1921), Prussian blue (PB) showed no obvious color changes after aging (see Figure 4b). However, it can be seen in Figure 4(c_2) that there is a significant intensity change in the UV absorption band. Therefore, the visual color maintenance may be attributed to the strong tinting ability and high saturation of the pigment.

After being mixed with the protective materials, the intensity of UV absorption band for each sample decreased because of the diluting effect of material powders (Figure 4(c_1,c_2)), which can be avoided to the greatest extent by adjusting the dosage and action mode of protection materials. For 1D-MCH- and 2D-MCH-treated samples, the type and position of the peaks in UV absorption bands remain unchanged, indicating that the protective material did not damage the color mechanism of the pigments. In contrast, commercial MgO had a serious effect on the color of LCY and PB. The UV curve changed obviously, and the color of LCY turned red, while PB almost completely faded. This kind of undesirable consequence may be due to the formation of more alkaline $Mg(OH)_2$ in the MgO dispersion, while both LCY and PB are very unstable under alkaline conditions and present different colors with different oxidation states. Therefore, for paper relics using alkali-sensitive pigments, it is safer to use weakly alkaline 1D-MCH and 2D-MCH than commercial MgO as deacidification agents.

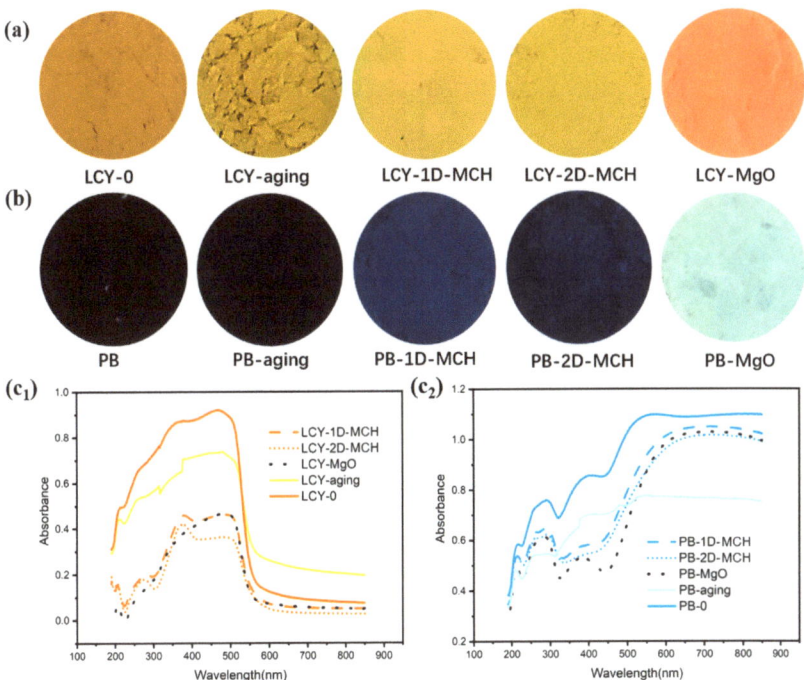

Figure 4. Photographs (a,b) and UV–Vis spectra of lead chrome yellow(c_1) and Prussian blue (c_2), mixed with 1D-MCH, 2D-MCH, and MgO.

2.3. Deacidification and Anti-Aging Properties

Acidification is one of the main diseases threatening the life of paper-based cultural relics, and thus, deacidification is crucial in extending the life span of paper relics. It should be noted that the goal of deacidification is not only to neutralize the acid already produced in the paper but also to deposit an alkaline substance to neutralize the acid that may be produced in the future, that is, to form a long-term alkaline reserve in cultural relics. Therefore, in addition to comparing the pH value before and after deacidification, the pH change in different treated paper samples during dry-heat aging was also tracked.

Before coating, the surface pH value of HMP was 4.07, reflecting that the paper was at a serious acidification level. After coating with three materials, respectively, the pH of the paper samples increased significantly. *The technical specifications and quality requirements for the restoration of ancient books* formulated by the National Museum of China stipulate that the paper after deacidification must be in the neutral or weakly alkaline region, which means the pH should between 7.5 and 10. Among the deacidification materials and the control groups used in this work, all three could keep the alkalinity of the paper samples within the desired range.

After 14 days of dry-heat aging, the pH value of each sample decreased to a certain extent, as Figure 5b displays. Although the results of the tests immediately after coating showed that the alkalinity of the 1D-MCH-coated samples were not highly alkaline, with a pH below 9, after 14 days of dry-heat aging, the pH of this group of specimens remained at a high level, which indicates that the material has a good durability in terms of deacidification.

In comparison, the 2D-MCH-treated samples showed a mediocre alkaline persistence, which may be related to the stability of the alkali carbonate. However, the pH of the commercial MgO-treated samples showed a large change from the first day of accelerated aging, decreasing from 9.54 to 8.27. After 14 days, the pH of this group of specimens was the lowest, indicating that it has a quick- rather than slow-release action.

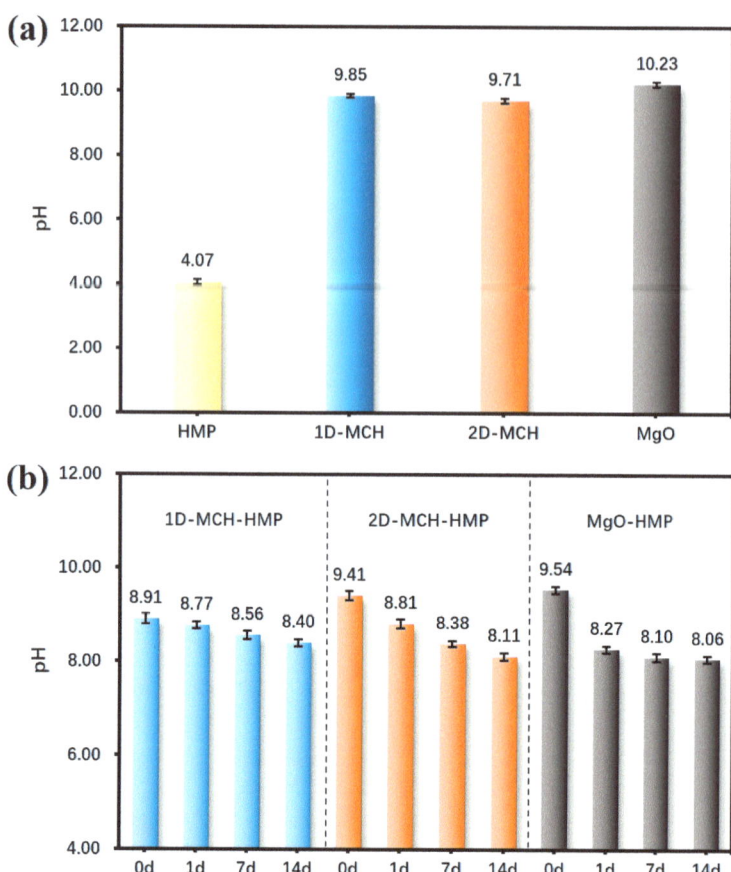

Figure 5. (a) The pH of HMP paper samples before coating and the dispersion of various materials (0.100 g material in 9.900 g 75 % ethanol); (b) the pH of coated samples undergoing dry-heat aging.

2.4. Strengthening Performance, Flame-Retardant Property, and Mechanism

Combining the results of safety test and deacidification experiments, it can be seen that although the alkalinity of 1D-MCH and 2D-MCH is lower than MgO, their long-term deacidification effect is better than that of MgO, and both of them are safe choices. The performance of the material's deacidification ability should be mainly attributed to its chemical composition, while the different structures of the two materials are expected to play a more differentiated role in the physical properties. Therefore, after the deacidification and anti-aging experiments, the different structural properties of 1D-MCH and 2D-MCH were further explored in the enhancement of paper mechanical properties.

The results of tearing resistance and folding endurance tests of the paper samples before and after coating show that both protective materials have an excellent strengthening effect on paper, especially in the enhancement of tearing resistance (Figure 6), which increased by ca. 128.4% (1D-MCH) and 64.2% (2D-MCH), respectively, compared with uncoated paper. It is obvious that 1D-MCH is better than 2D-MCH in improving the mechanical properties of paper relics, although the latter one can already achieve good results. Further, the test results of the ultimate oxygen index show that both types of low-dimensional materials with different morphologies can have a certain flame retardant effect on HMP (Figure 6c). The limit oxygen index of 1D-MCH is higher than that of 2D-MCH, reaching 22.2%, which is higher than the oxygen content in the air (about 21%, marked with the red dashed line in Figure 6c). This indicates that HMP loaded with 1D-MCH is difficult to burn stably in air, and the material can achieve good flame retardant protection effect.

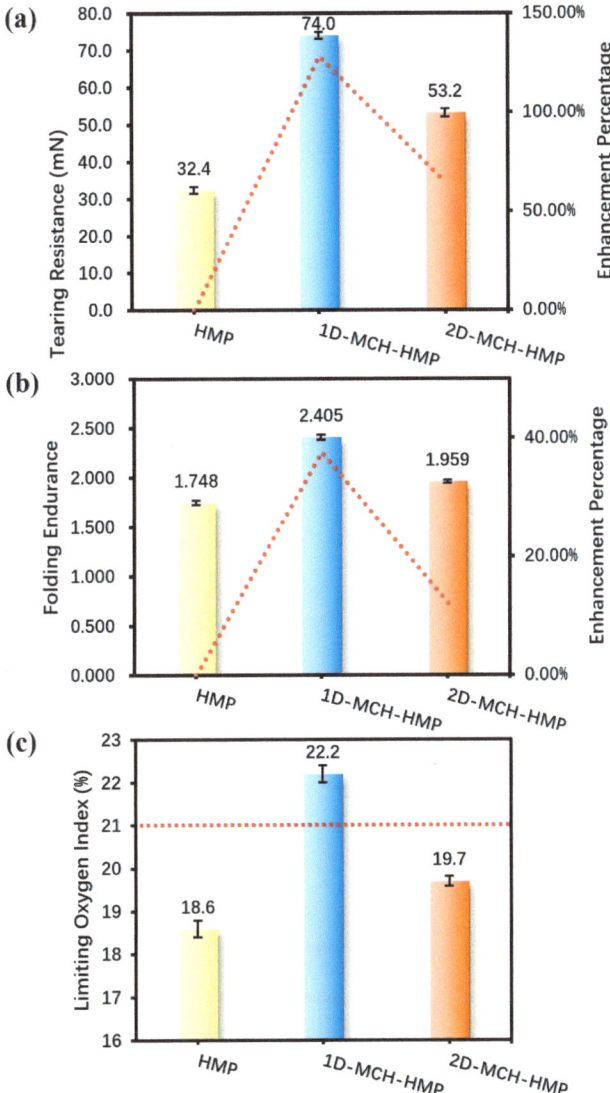

Figure 6. (a) Tearing resistance and enhancement percentage (red line) of different samples; (b) folding endurance and enhancement percentage (red line) of different samples; (c) limiting oxygen index of different samples.

The SEM images and related element analysis results confirm that through the coating procedure, the protective materials can deposit uniformly in the paper fibers and could serve to fill the voids between fibers (Figure 7a–d), even when the fibers of HMP were flattened during its production process. The good coating condition provides the possibility for low-dimensional materials to play the role of strengthening and flame retardancy.

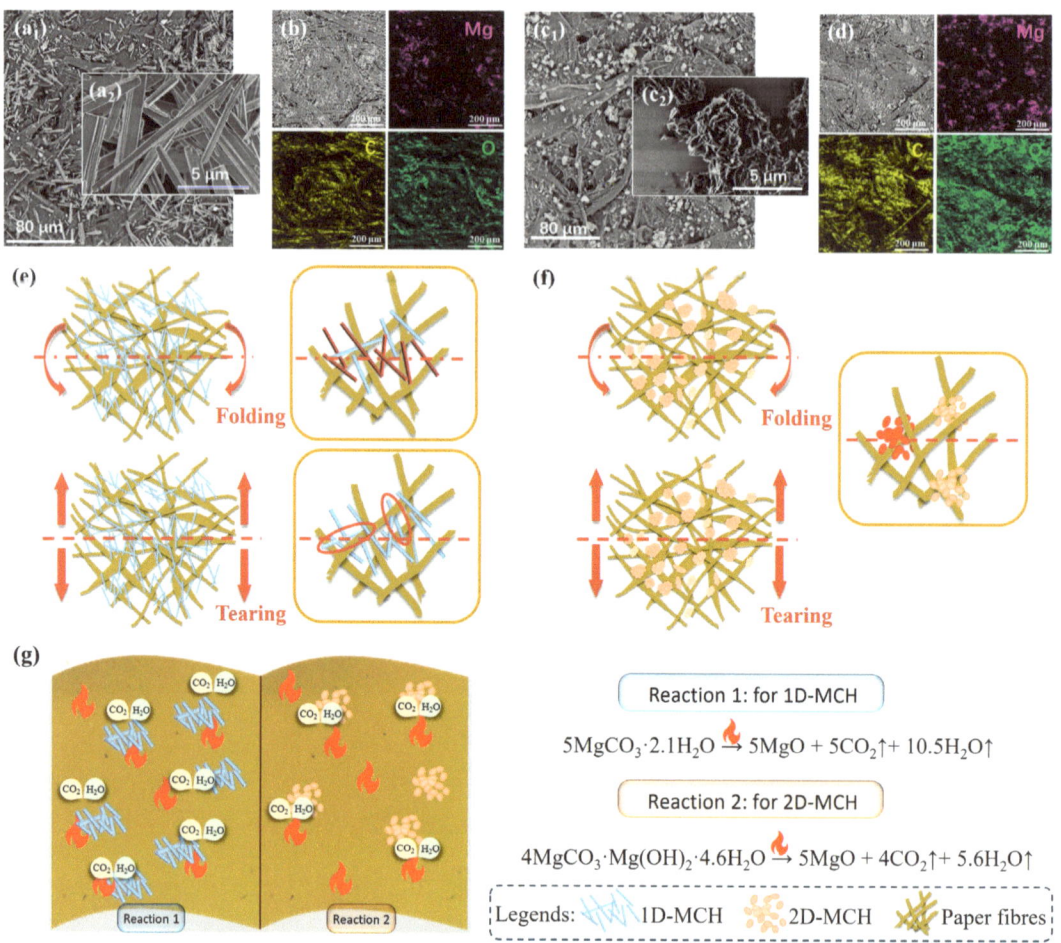

Figure 7. (a_1,c_1) SEM images of paper samples coated by 1D-MCH and 2D-MCH; (a_2,c_2) FESEM images of 1D-MCH and 2D-MCH; (**b**,**d**) SEM images and related Mg, C, and O mapping images of paper samples coated by 1D-MCH and 2D-MCH; (**e**–**g**) schematic diagrams of 1D-MCH and 2D-MCH deposition in paper fibers and their reinforcement and flame retardant mechanisms.

Further analysis of the protective mechanisms of two different low-dimensional materials in HMP reveals that the reinforcement effect of materials is not only related to the filling density but also depends on the mechanical strength of the material itself and the mode of action between protection materials and paper fibers.

As a kind of single crystal fiber, one-dimensional whisker material is very small in diameter, so it is difficult to accommodate internal defects in large crystals, and thus, its strength is close to the theoretical value of complete crystals. As illustrated in Figure 7e, for 1D-coated samples, the deposition of 1D-MCH not only fills the fiber gaps and increases their density, but also provides some support when the paper is bent. Thanks to the high strength of the whiskers, this one-dimensional deposit can counteract part of the impact of external forces to protect the paper fibers, thus significantly improving their folding resistance. In addition, when the paper is subjected to tearing tension, the one-dimensional material can act as a bridge between the paper fibers, enhancing its resistance to tearing.

In contrast, the filling of two-dimensional materials, while also strengthening the fragile paper, is not as effective as one-dimensional materials. Although theoretically, the flexibility brought by the nanosheets' extremely thin size can help the material to fill in the gaps between the paper fibers, the SEM image and related element analysis results revealed that 2D-MCH failed to be distributed as layers in the paper. Instead, the nanosheets formed flower-like aggregates and presented in fibers in a filling or adhesion way, see Figure 7c.

Meanwhile, the flame retardancy of the material is related to its composition and thermal stability. Both 1D-MCH and 2D-MCH belong to hydrated magnesium carbonate materials, which can decompose after being heated and release water and carbon dioxide to delay combustion, as shown in Figure 7g. It can be seen from the TGA results (Figure 2c,d) that the ratio of water and carbon dioxide released by 1D-MCH after complete decomposition is higher, and the loss of its bound water begins at 180 °C, while the decomposition of 2D-MCH begins below 100 °C. The relatively high thermal stability and high levels of water and carbon dioxide in the composition are both reasons why 1D-MCH is able to show better flame retardancy.

3. Experimental Section

3.1. Reagents and Materials

Magnesium nitrate hexahydrate ($Mg(NO_3)_2 \cdot 6H_2O$, analytical reagent (AR), Shanghai Chemical Co., Shanghai, China) and sodium bicarbonate ($NaHCO_3$, AR, Sinopharm, Beijing, China) were used to synthesize products with different micro-morphologies. For protective tests, naturally aged handmade wood-pulp paper from a book published in 1949 (denoted as HMP) were used as paper samples. Lead chrome yellow ($PbCrO_4$, AR, Shanghai Chemical Co., Shanghai, China, denoted as LYC) and Prussian blue ($Fe_4[Fe-(CN)_6]_3$, AR, Shanghai Chemical Co., Shanghai, China, denoted as PB) were selected to test the safety of different protective materials. A commercial metallic oxide reagent, magnesium oxide (MgO, AR, Sinopharm, Beijing, China), was regarded as the reference sample for both deacidification and anti-discoloration tests. Deionized water and anhydrous ethanol were used throughout.

3.2. Methods

3.2.1. Preparation of Low-Dimensional Materials

This paper adopts a convenient and energy-efficient supersaturation adjust method to obtain 1D and 2D materials. What Wei-Hong Lai and co-workers have previously done is to judiciously tune the reaction system by adjusting the electrolytic dissociation (α) of the precursors and the supersaturation (S) of the solution to favor the growth of specific morphologies, without the need for any additives [23]. Combining the methods reported in the literature, we used $Mg(NO_3)_2$ as the magnesium ion providing agent, and controlled the ionization degree of the reactants in the solution by the weak electrolyte $NaHCO_3$, on the basis of which the degree of supersaturation of the solution was regulated in order to obtain the products with different microcosmography.

For the 1D product, 0.065 M $Mg(NO_3)_2 \cdot 6H_2O$ was dissolved in 30 mL deionized water, heated to 80 °C, and stirred for 30 min at the speed of 300 rpm to obtain an even solution. Then, 0.198 M $NaHCO_3$ was added into the solution as quickly as possible. The products of the reaction were collected by centrifugation after 5 min, washed with water three times and freeze-dried for three days. The supersaturation of the system was adjusted to obtain the 2D product, in which 0.130 M $Mg(NO_3)_2 \cdot 6H_2O$ and 0.600 M $NaHCO_3$ were used, while the other steps remained unchanged.

3.2.2. Coating Procedure

Protective materials were deposited on paper samples through a dip-coating method. The paper samples were cut into several sizes of specimens (2×2 cm^2 for accelerated-aging, 5.0×6.5 cm^2 for tearing test, 1.5×10 cm^2 for folding endurance test, and 14×5.2 cm^2 for flame retardancy test). Each paper sheet went through a round of 5 min immerse in 10 mL preconfigured protective suspension (0.100 g material in 9.900 g 75% ethanol) and 5 min naturally drying, performed 5 times in total. And then all the coated papers were dried under a vacuum condition at 25 °C for 24 h for subsequent testing.

3.2.3. Dry-Heat Aging

The aging of paper samples was accelerated through an artificial dry-heat process by keeping them in a thermostatic chamber at 105 °C with air circulation (ISO 5630–1:1991) [31]. The paper samples were stabilized at 25 °C for at least 24 h before and after aging (ISO 187–1990) [32].

3.2.4. Safety Test

To evaluate the safety of protection materials, the contact angle and chromaticity changes in the pattern were tested. Two alkali-sensitive pigments, PB and LCY, were selected for the safety test. In this experiment, 3 mL pigment dispersion (1 wt%) was mixed with 7 mL protective suspension. After being shaken for 5 min and heated in an oven at 80 °C for 24 h, the mixture was filtered and dried for the UV-Vis test.

3.2.5. Protection Effect

The protection performance of materials was evaluated via deacidification and the enhancement of mechanical properties of samples after coating. The pH values of paper samples coated with different materials were compared to evaluate the deacidification effect of the materials. Then, on the 1st, 7th, and 14th days of accelerated aging, the paper samples were extracted for pH testing to judge the anti-aging properties of the protective agents. In terms of the enhancement effect of mechanical properties, the tearing resistance, folding endurance, and flame retardancy of the paper samples before and after coating were compared.

3.3. Characterization

Powder X-ray diffraction (XRD) was tested on a Bruker D2-Advanced diffractometer (Bruker Corporation, Billerica, MA, USA) from 5 to 80° with a scanning rate of 5° min^{-1}. The related element analysis, morphologies, and structures were investigated by scanning electron microscopy (SEM, Phenom Prox, PhenomWorld, Eindhoven, The Netherlands), field emission scanning electron microscopy (FESEM, Nova NanoSEM 450, Thermo Fisher Scientific, Eugene, OR, USA), transmission electron microscopy (TEM, JEM 2010, JEOL Ltd., Tokyo, Japan), and field emission transmission electron microscopy (FETEM, Tecnai G2 F20 S-Twin, Thermo Fisher Scientific, Eugene, OR, USA). A thermogravimetric analysis (TGA) was performed on a SDT Q600 thermal analysis instrument (TA Instruments, Columbus, OH, USA) with a heating rate of 10 °C/min under air flow (100 mL/min). The UV-vis absorption spectra was obtained by a Lambda 650 spectrophotometer (PERKIN ELMER, Hopkinton, MA, USA). For the properties of the paper samples, the static contact angles were measured on an OCA15+ contact angle system (Dataphysics, Falkenhagen, Germany) at ambient temperature using a water drop volume of 5 µL, the pH values of paper samples were measured using a paper pH meter (HANNA HI 99171, HANNA Instruments, Sliema, Malta) with a planar electrode, and the chroma changes in the paper samples were analyzed using an NR10QC colorimeter (Konica Minolta, Tokyo, Japan). In terms of mechanical properties, the folding endurance was performed by referring to ISO 5626:1993 [33], and the tearing resistance was conducted according to ISO 1974:1990 [34] with a tension of 1.5 N. The flame retardancy of the sample was tested according to GB/T 2406.2 [35].

4. Conclusions

In this paper, we report the use of a facile synthesis strategy to obtain well-formed 1D and 2D MCHs, which were applied in experiments for the protection of paper-based cultural relics. The safety performance of the two materials was tested according to the special requirements of cultural relic protection, and the results showed that both materials did not lead to significant discoloration of paper, nor did they cause significant changes in the color of alkali-sensitive pigments. The results of the deacidification and aging experiments show that the deacidification effect of both materials can meet the requirements for the protection of paper cultural relics, and successfully form a long-term alkaline reserve in the paper samples, with a certain degree of anti-aging effect. In addition, thanks to its own structural properties, the 1D-MCH performs well in enhancing the mechanical strength of paper, and can bring about a flame-retardant effect, realizing the combination of safe deacidification protection and effective enhancement. In this study, the mild deacidification and anti-aging protection of paper cultural relics were realized from the synthetic modulation of material micromorphology, and the strength-enhancing effects of low-dimensional materials with different micromorphologies and their mechanisms were explored, which provided a new vision for further research on paper cultural relic protection materials.

Author Contributions: Conceptualization, X.J., H.Z. and Y.T.; methodology, Y.W. and Z.Z.; data curation, P.L. and J.W.; writing—original draft preparation, Z.Z. and Y.W.; writing—review and editing, H.Z. and J.W.; project administration, Y.T.; funding acquisition, P.L., X.J. and H.Z. All authors have read and agreed to the published version of the manuscript.

Funding: This work was supported by the National Natural Science Foundation of China (Grant No. 22175040), and the Foundation of State Key Laboratory of Biobased Material and Green Papermaking, Qilu University of Technology, Shandong Academy of Sciences (Grant Nos. GZKF202109, GZKF202210).

Institutional Review Board Statement: Not applicable.

Informed Consent Statement: Not applicable.

Data Availability Statement: Data are contained within the article.

Conflicts of Interest: The authors declare no conflicts of interest.

References

1. Banks, P.N. Some problems in book conservation. *Libr. Resour. Tech. Serv.* **1968**, *12*, 330–338.
2. Bruckle, I. Bleaching in Paper Production versus Conservation. *Restaur.-Int. J. Preserv. Libr. Arch. Mater.* **2009**, *30*, 280–293. [CrossRef]
3. Lanzón, M.; Madrid, J.A.; Martínez-Arredondo, A.; Mónaco, S. Use of diluted $Ca(OH)_2$ suspensions and their transformation into nanostructured $CaCO_3$ coatings: A case study in strengthening heritage materials (stucco, adobe and stone). *Appl. Surf. Sci.* **2017**, *424*, 20–27. [CrossRef]
4. Weng, J.; Zhang, X.; Jia, M.; Zhang, J. Deacidification of aged papers using dispersion of $Ca(OH)_2$ nanoparticles in subcritical 1,1,1,2-tetrafluoroethane (R134a). *J. Cult. Herit.* **2019**, *37*, 137–147. [CrossRef]
5. Malešič, J.; Kadivec, M.; Kunaver, M.; Skalar, T.; Cigić, I.K. Nano calcium carbonate versus nano calcium hydroxide in alcohols as a deacidification medium for lignocellulosic paper. *Herit. Sci.* **2019**, *7*, 50. [CrossRef]
6. Lisuzzo, L.; Cavallaro, G.; Milioto, S.; Lazzara, G. Halloysite nanotubes filled with MgO for paper reinforcement and deacidification. *Appl. Clay Sci.* **2021**, *213*, 106231. [CrossRef]
7. Castillo, I.F.; De Matteis, L.; Marquina, C.; Guillén, E.G.; de la Fuente, J.M.; Mitchell, S.G. Protection of 18th century paper using antimicrobial nano-magnesium oxide. *Int. Biodeterior. Biodegrad.* **2019**, *141*, 79–86. [CrossRef]
8. Huang, J.; Liang, G.; Lu, G.; Zhang, J. Conservation of acidic papers using a dispersion of oleic acid-modified MgO nanoparticles in a non-polar solvent. *J. Cult. Herit.* **2018**, *34*, 61–68. [CrossRef]
9. Alexopoulou, I.; Zervos, S. Paper conservation methods: An international survey. *J. Cult. Herit.* **2016**, *21*, 922–930. [CrossRef]
10. Baty, J.W.; Maitland, C.L.; Minter, W.; Hubbe, M.A.; Jordanmowery, S.K. Deacidification for the conservation and preservation of paper-based works: A Review. *Bioresources* **2010**, *5*, 1955–2023. [CrossRef]
11. Tian, P.; Shi, M.; Hou, J.; Fu, P. Cellulose-Graphene Bifunctional Paper Conservation Materials: For Reinforcement and UV Aging Protection. *Coatings* **2023**, *13*, 443. [CrossRef]

12. Wu, C.; Jin, C.; Zhang, W.; Cui, X.; Li, C.; Zhu, Z.; Liu, Y.; Liu, P.; Zhang, H.; Zhang, H.; et al. Exploration on the regulation of writing performance through Pickering emulsions-coated handmade Xuan paper. *Cellulose* **2024**, *31*, 1295–1309. [CrossRef]
13. Zhao, H.; Liu, P.; Huang, Y.; Zhang, H. Nanocomposites composed of modified natural polymer and inorganic nanomaterial for safe, high-efficiency, multifunctional protection of paper-based relics. *Sci. China-Technol. Sci.* **2023**, *66*, 2225–2236. [CrossRef]
14. Wu, L.; Jin, Z.L.; Zhang, Z.H. Application and synthesis of inorganic whisker materials. *Prog. Chem.* **2003**, *15*, 264–274.
15. Chen, X.; Qian, X.; An, X. Preparation of flame-retardant paper using magnesium salt whiskers as filler. *Appita J.* **2012**, *65*, 87–92.
16. Chen, X.; Qian, X.; An, X. Using calcium carbonate whiskers as papermaking filler. *Bioresources* **2011**, *6*, 2435–2447. [CrossRef]
17. Zhang, C.; Huang, Y.; Zhao, H.; Zhang, H.; Ye, Z.; Liu, P.; Zhang, Y.; Tang, Y. One-Pot Exfoliation and Functionalization of Zeolite Nanosheets for Protection of Paper-Based Relics. *ACS Appl. Nano Mater.* **2021**, *4*, 10645–10656. [CrossRef]
18. Zhang, C.; Huang, Y.; Zhang, H.; Ye, Z.; Liu, P.; Wang, S.; Zhang, Y.; Tang, Y. Selectively Functionalized Zeolite NaY Composite Materials for High-Efficiency Multiple Protection of Paper Relics. *Ind. Eng. Chem. Res.* **2020**, *59*, 11196–11205. [CrossRef]
19. Zhang, H.; Zhang, C.; Ye, Z.; Wang, S.; Tang, Y. Alkali-exchanged Y zeolites as superior deacidifying protective materials for paper relics: Effects of accessibility and strength of basic sites. *Microporous Mesoporous Mater.* **2020**, *293*, 109786. [CrossRef]
20. Wang, S.; Yang, X.; Li, Y.; Gao, B.; Jin, S.; Yu, R.; Zhang, Y.; Tang, Y. Colloidal magnesium hydroxide Nanoflake: One-Step Surfactant-Assisted preparation and Paper-Based relics protection with Long-Term Anti-Acidification and Flame-Retardancy. *J. Colloid Interface Sci.* **2022**, *607*, 992–1004. [CrossRef]
21. Lin, H.-X.; Lei, Z.-C.; Jiang, Z.-Y.; Hou, C.-P.; Liu, D.-Y.; Xu, M.-M.; Tian, Z.-Q.; Xie, Z.-X. Supersaturation-Dependent Surface Structure Evolution: From Ionic, Molecular to Metallic Micro/Nanocrystals. *J. Am. Chem. Soc.* **2013**, *135*, 9311–9314. [CrossRef] [PubMed]
22. Lovette, M.A.; Doherty, M.F. Predictive Modeling of Supersaturation-Dependent Crystal Shapes. *Cryst. Growth Des.* **2012**, *12*, 656–669. [CrossRef]
23. Lai, W.-H.; Wang, Y.-X.; Wang, Y.; Wu, M.; Wang, J.-Z.; Liu, H.-K.; Chou, S.-L.; Chen, J.; Dou, S.-X. Morphology tuning of inorganic nanomaterials grown by precipitation through control of electrolytic dissociation and supersaturation. *Nat. Chem.* **2019**, *11*, 695–701. [CrossRef]
24. Mergelsberg, S.T.; Kerisit, S.N.; Ilton, E.S.; Qafoku, O.; Thompson, C.J.; Loring, J.S. Low temperature and limited water activity reveal a pathway to magnesite via amorphous magnesium carbonate. *Chem. Commun.* **2020**, *56*, 12154–12157. [CrossRef]
25. Rheinheimer, V.; Unluer, C.; Liu, J.; Ruan, S.; Pan, J.; Monteiro, P.J.M. XPS Study on the Stability and Transformation of Hydrate and Carbonate Phases within MgO Systems. *Materials* **2017**, *10*, 75. [CrossRef]
26. Rincke, C.; Schmidt, H.; Voigt, W. A new hydrate of magnesium carbonate, $MgCO_3 \cdot 6H_2O$. *Acta Crystallogr. Sect. C-Struct. Chem.* **2020**, *76*, 244–249. [CrossRef] [PubMed]
27. Zhang, Z.; Zheng, Y.; Ni, Y.; Liu, Z.; Chen, J.; Liang, X. Temperature- and pH-dependent morphology and FT-IR analysis of magnesium carbonate hydrates. *J. Phys. Chem. B* **2006**, *110*, 12969–12973. [CrossRef]
28. Chen, Q.; Hui, T.; Sun, H.; Peng, T.; Ding, W. Synthesis of magnesium carbonate hydrate from natural talc. *Open Chem.* **2020**, *18*, 951–961. [CrossRef]
29. Zheng, Y.; Dang, L.; Zhang, Z. Influence of Stirring Time on the Morphology and Composition of Magnesium Carbonate Hydrates. *Fine Chem.* **2007**, *24*, 835–838, 869.
30. Miliani, C.; Monico, L.; Melo, M.J.; Fantacci, S.; Angelin, E.M.; Romani, A.; Janssens, K. Photochemistry of Artists' Dyes and Pigments: Towards Better Understanding and Prevention of Colour Change in Works of Art. *Angew. Chem.-Int. Ed.* **2018**, *57*, 7324–7334. [CrossRef]
31. *ISO 5630–1:1991*; Paper and Board—Accelerated Ageing—Part 1: Dry Heat Treatment at 105 Degrees C. ISO: Geneva, Switzerland, 1991.
32. *ISO 187:1990(en)*; Paper, Board and Pulps—Standard Atmosphere for Conditioning and Testing and Procedure for Monitoring the Atmosphere and Conditioning of Samples. ISO: Geneva, Switzerland, 1990.
33. *ISO 5626:1993*; Paper—Determination of Folding Endurance. ISO: Geneva, Switzerland, 1993.
34. *ISO 1974:1990*; Paper—Determination of Tearing Resistance (Elmendorf Method). ISO: Geneva, Switzerland, 1990.
35. *GB/T 2406.2-2009*; Plastics—Determination of Burning Behaviour by Oxygen Index—Part 2: Ambient-Temperature Test. Standardization Administration of China: Beijing, China, 2009.

Disclaimer/Publisher's Note: The statements, opinions and data contained in all publications are solely those of the individual author(s) and contributor(s) and not of MDPI and/or the editor(s). MDPI and/or the editor(s) disclaim responsibility for any injury to people or property resulting from any ideas, methods, instructions or products referred to in the content.

Multi-Functional Repair and Long-Term Preservation of Paper Relics by Nano-MgO with Aminosilaned Bacterial Cellulose

Hongyan Mou [1,*], Ting Wu [1], Xingxiang Ji [2], Hongjie Zhang [3], Xiao Wu [1,*] and Huiming Fan [1]

1. State Key Laboratory of Pulp and Paper Engineering, School of Light Industry and Engineering, South China University of Technology, Guangzhou 510640, China
2. Key Laboratory of Pulp and Paper Science & Technology of Ministry of Education, Qilu University of Technology (Shandong Academy of Sciences), Jinan 250353, China; xxjt78@163.com
3. National Engineering Laboratory for Pulp and Paper, Beijing 100102, China
* Correspondence: fehymou@scut.edu.cn (H.M.); m13719469536@163.com (X.W.)

Abstract: Paper relics, as carrieres of historical civilization's records and inheritance, could be severely acidic and brittle over time. In this study, the multi-functional dispersion of nanometer magnesium oxide (MgO) carried by 3-aminopropyl triethoxysilane-modified bacterial cellulose (KH550-BC) was applied in the impregnation process to repair aged paper, aiming at solving the key problems of anti-acid and strength recovery in the protection of ancient books. The KH550-BC/MgO treatment demonstrated enhanced functional efficacy in repairing aged paper, attributed to the homogeneous and stable distribution of MgO within the nanofibers of BC networks, with minimal impact on the paper's wettability and color. Furthermore, the treatment facilitated the formation of adequate alkali reserves and hydrogen bonding, resulting in superior anti-aging properties in the treated paper during prolonged preservation. Even after 30 days of hygrothermal aging tests, the paper repaired by KH550-BC/MgO was still in a gently alkaline environment (pH was about 7.56), alongside a 32.18% elevation compared to the untreated paper regarding the tear index. The results of this work indicate that KH550-BC/MgO is an effective reinforcement material for improving the long-term restoration of ancient books.

Keywords: paper relics; conservation; bacterial cellulose; nano-MgO; aging resistance

1. Introduction

Paper literature has an incalculable significance and serves as a vital conduit for historical civilization in many eras and nations [1]. Affected by the pulp and paper technology and the harsh storage environment, a large number of documents and books are irreversibly acidified and aged, seriously affecting the service and storage life [2–5]. In recent years, countries around the world have begun to pay attention to and explore the research topic of the protection and restoration of ancient books. The previous repair process typically involved a series of step-by-step processes, including deacidification, reinforcement, and antibacterial and mold prevention [6,7]. These processes were complex and inefficient and could easily cause secondary damage to the paper. Therefore, multi-functional methods for restoring ancient books have become a key research direction [8–11].

Bacterial cellulose (BC), as a green natural polymer material, has an ultra-fine three-dimensional porous fiber structure and excellent mechanical properties [12,13], offering great prospects in the reinforcement and protection of paper documents [14–16]. In our previous study, BC-repaired paper exhibited a high folding endurance of 28 times, indicating a promising application of BC as a restoration material [17]. However, the long-term stability of BC for the preservation of ancient books is limited by its hydrophilicity and the fact that BC cannot be deacidified [18]. The unique chemical composition of BC, which is rich in hydroxyl groups, offers a promising avenue for the development of new materials with more functionalities [19]. For instance, mineralized BC had been successfully employed in the deacidification of paper-based materials, while simultaneously imparting paper with exceptional flame retardant properties [20]. Another reliable approach for the repurposing of BC is amino silanization [21–23]; the results of our latest research indicated that this process improves the interface bonding between BC and aged paper toward higher mechanical properties of the treated paper [24].

Furthermore, adequate alkali reserves are usually considered as a guarantee of long-term acid resistance in paper [25]. Recently, alkaline nanoparticles have been the favored deacidifying materials because of their smaller particle size and better permeability, which helps to further extend the durability of paper [26,27]. From a more positive perspective, nano-magnesia is regarded by conservation scientists as a preferred deacidifier among magnesium derivatives, primarily due to its milder alkalinity compared to magnesium hydroxide. However, an inherent limitation of nano-MgO is its tendency to settle out in the solvent because of its poor dispersion stability. Oleic-acid-modified nano-MgO was used to promote its dispersion stability in a cyclohexane non-polar solvent for paper deacidification [28]. He et al. [29] also created a stable organic covering that allowed for the uniform deposition of nano-MgO particles on paper, based on the co-dispersion of trimethylsilyl cellulose (TMSC) and isopropyl alcohol (IPA). Our theory posits that the long-chain network structure of KH550-BC could impede the agglomeration of nano-MgO particles through steric hindrance while enhancing the alkalinity of KH550-BC, which is beneficial for the long-term stability of KH550-BC/MgO-reinforced paper.

Herein, a nano-composite dispersion of KH550-BC and MgO was prepared and developed by adhering to the principle of "repairing old as before". The repair effect and anti-aging properties of aged paper were improved using KH550-BC/MgO, and the process was comprehensively investigated. The anti-aging mechanism of KH550-BC/MgO to aged paper was studied via ATR-Fourier infrared spectroscopy (ATR-FTIR) and X-ray diffraction spectroscopy (XRD), and the distribution and deposition of KH550-BC/MgO in paper were analyzed through field-emission scanning electron microscopy (FESEM) as a supplement for evaluating the long-term stability of KH550-BC/MgO-reinforced aged paper.

2. Results and Discussion

2.1. Effect of Loading Nano-MgO in KH550-BC on Deacidification

In the KH550-BC/MgO system, the addition of nano-MgO should give the paper enough alkali storage to extend the service life of the paper, but pH and alkali storage that are too high would cause the alkaline degradation of the paper fiber and yellowing of the paper after deacidification [30–32]. To reduce these undesirable changes, we investigated the effect of nano-MgO at varying concentrations, and the results were compared with KH550-BC.

As shown in Table 1, with an increase in MgO content in the KH550-BC/MgO system, the pH value of the repaired paper significantly increased, and the alkali reserves, as the key factor for the anti-acidification of paper, exceeded 0.30 mol/kg. In images of paper repaired by KH550 loading different contents of MgO, the MgO at 0.1% could distribute on paper evenly, as shown in Figure 1b. However, at concentrations above 0.1% (as the green arrows pointed), MgO with a high specific surface energy was prone to agglomeration and deposition on the paper surface. As a consequence, 0.1% MgO was chosen to be appropriate for paper restoration in accordance with the lack of interference with paper font ink. At this point, the grammage of the paper was increased by 3.25 g/m^2, and its alkali reserve was 0.36 mol/kg, which was 41.18% higher than that of KH550-BC repair alone (0.26 mol/kg).

Table 1. The influence of MgO contents in the KH550-BC/MgO system on the pH and alkali reserve of aged paper.

Samples	Mass Fraction of MgO/%	pH	Alkali Reserve/mol kg^{-1}
1	0	7.96	0.26
2	0.1	9.42	0.36
3	0.2	9.98	0.53
4	0.3	10.21	0.82

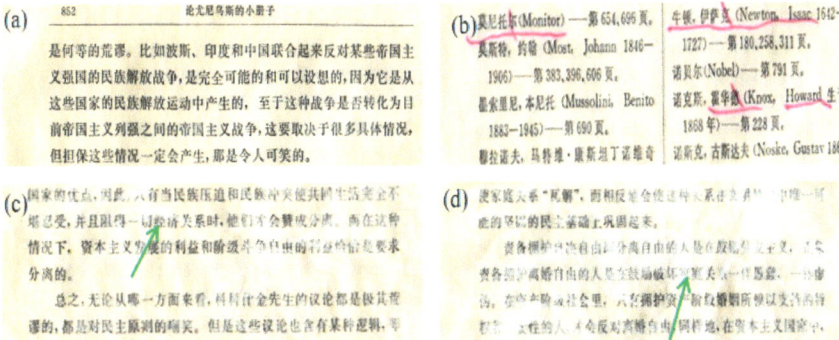

Figure 1. The influence of MgO contents in the KH550-BC/MgO system on the appearance of the paper samples, wherein (**a**) represented the paper repaired by KH550-BC, (**b**–**d**) represented the paper repaired by KH550-BC/MgO, in which the MgO content was (**b**) 0.1%, (**c**) 0.2%, or (**d**) 0.3%.

2.2. Strengthening Effect of KH550-BC/MgO on Aged Paper

While KH550-BC/MgO had the better deacidification effect on aged paper (Table 1), the strengthening effects of KH550-BC/MgO on aged paper were also maintained at a high level (Table 2). P2 outperformed the reference sample (UP) in terms of the tear index, tensile index, and folding endurance, with values of 6.60 mN m^2/g, 35.44 N m/g, and 19 times, respectively. In addition, the impregnation treatment with KH550-BC/MgO resulted in a 13.91% increase in the zero-span tensile strength, while the tensile stress also increased by 88.60%. It could be indicated that KH550-BC/MgO has significantly improved the strength of the individual fibers and the tensile performance. All of these phenomena may be induced by the hydrogen bonds between modified bacterial cellulose being opened, exposing the free hydroxyl groups on its molecules, which form more hydrogen bonds with paper fibers, thereby increasing the binding force between fibers of aged paper. Meanwhile, some BC was adsorbed on the surface of fibers, interweaving and winding to form a network structure, thereby increasing the strength of each paper fiber [17]. This hypothesis will be verified by subsequent analysis.

Table 2. Comparison of the aged paper repairing effects of two reinforcement systems.

Sample	pH	Alkali Reserve (mol/kg)	Tear Index (mN·m^2/g)	Tensile Index (N·m/g)	Folding Endurance (Times)	Zero-Span Tensile Strength (kN/m)	Tensile Stress (MPa)
UP	6.45	0	3.48	18.70	4.50	68.30	14.27
P1	7.96	0.26	6.65	36.15	19.80	78.60	28.02
The increased ratio (%)	23.41	—	91.09	93.32	340	15.08	96.36
P2	9.42	0.36	6.60	35.44	19	77.80	27.29
The Increased ratio (%)	46.05	—	89.66	89.52	322.20	13.91	88.60

2.3. Wettability Analysis of Repaired Paper

Following the repair with KH550-BC/MgO, Figure 2 shows that the contact angle between the paper and water was 118°, which was nearly identical to that of the base paper UP (119°). The hydrophobicity of UP was correlated with the chemical agent used in the papermaking process. In order to prevent the paper from being damaged by wetting, sizing agents were added to enhance the paper's resistance to liquid penetration and diffusion ability [33–35]. However, as shown in Figure 2b, the addition of BC, which contained a large number of hydrophilic hydroxyl groups, rendered the paper susceptible to deterioration when exposed to humid conditions over an extended duration following the repair process, due to the intrinsic hygroscopic nature of BC [36]. This weakness was improved by modifying BC with the inclusion of nano-MgO [13]. Specifically,, the surface effect of nano-MgO counterbalanced the hydrophilic effect of BC, resulting in a rougher surface and decreased porosity of the paper. As a result, the repaired paper was able to retain its hydrophobicity, ensuring its preservation in various environments for extended periods of time.

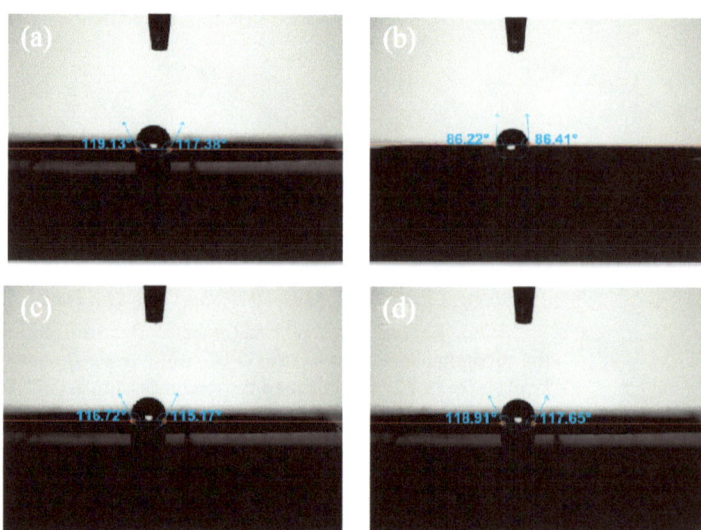

Figure 2. The water contact angles of (**a**) the untreated paper and (**b**–**d**) the paper repaired by BC, KH550-BC, and KH550-BC/MgO.

Furthermore, as illustrated in Figure 3, the SEM images of the paper prior to and following the KH550-BC/MgO repair process confirmed that KH550-BC/MgO effectively covered the fiber surface and bridged between the fiber, resulting in a notable enhancement of the paper's mechanical properties and a reduction in its porosity (Figure 3a,b). At 5000× magnification, the stable dispersion and uniform adsorption of nano-sized MgO in the three-dimensional fiber network of BC were clearly observed in Figure 3c, demonstrating the stable deacidification of KH550-BC/MgO and its potential for durability.

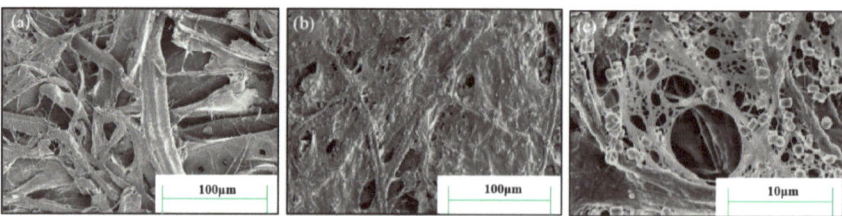

Figure 3. FESEM images of (**a**) the untreated paper and (**b**,**c**) the paper repaired by KH550-BC/MgO, which were magnified by (**b**) 500× and (**c**) 5000×.

2.4. Aging Resistance and Stability Studies of Repaired Paper

To further confirm the fact that KH550-BC/MgO improves the anti-aging properties of paper, artificial aging tests were carried out on aged paper before and after repair. The changes in the physical strength, acid–base property, and color appearance of all of the repaired paper samples during artificial aging were meticulously analyzed.

2.4.1. Mechanical Properties Change in Repaired Paper during Aging

A thorough investigation was conducted into the variations in the strength of the paper reinforced with KH550-BC and KH550-BC/MgO as they aged for different days. The results are presented in Figure 4. Given that the untreated paper had an acceptable initial strength (tear index of 3.48 mN·m^2/g, tensile index of 18.70 N·m/g, folding endurance of 4.5 times), there was no discernible decrease in the strength of any of the repaired paper samples after 3 days of artificial aging.

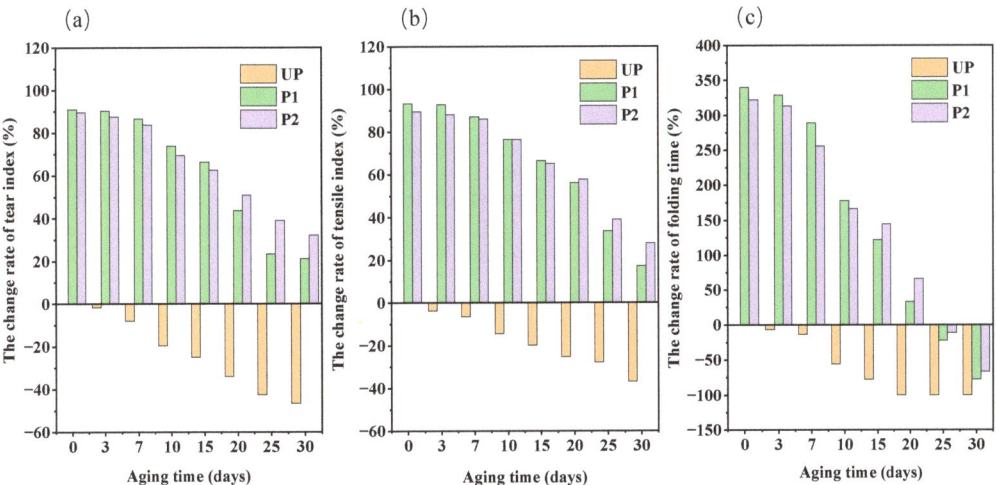

Figure 4. The change rate of the paper samples strength properties with aging time, in which (**a**) represented the tear index, (**b**) represented the tensile index, and (**c**) represented the folding endurance.

As illustrated in Figure 4a, during the aging period from 3 d to 15 d, the tear index of the untreated paper (UP) dropped sharply by 25.00%. This decline was attributed to the accumulation of acid rather than its consumption over time, leading to the constant acid degradation of cellulose in UP. Relevant research has also demonstrated a positive correlation between the concentration of H$^+$ and the rate of cellulose hydrolysis [37,38]. In comparison to the initial tear index of UP, the tear index of P1 and P2 after 15 days of aging was 66.38% and 62.64% higher, respectively. The enhancement effect of KH550-BC/MgO was not only reflected in the significant improvement of the interfacial adhesion between the BC and the aged paper, but more importantly, it provided enough of an alkali reserve for the repaired paper to continue to consume acidic substances produced in high-temperature and high-humidity environments. Therefore even after 30 d of aging, the tear index of P2 (4.60 mN·m^2/g) was higher than that of P1(4.22 mN·m^2/g), and it was still 32.18% higher than that of UP. The MgO present in KH550-BC acts as a strengthening agent, significantly enhancing the paper's anti-aging ability and prolonging its conservation duration.

The tensile index changes in all paper samples are shown in Figure 4b. On the 15th day of aging, the tensile index of UP decreased by 19.84%. Simultaneously, the paper enhanced by KH550-BC and KH550-BC/MgO dispersion showed a slight decrease, yet still retained a higher tensile index than UP, at 66.42% and 65.24%, respectively. The tensile index of P2 progressively surpassed that of P1 during the aging process of 15–20 days, and its reduction was the least during the whole aging period, which also indicates that the paper samples treated with KH550-BC/MgO had a good aging resistance stability [30].

Nonetheless, the changes in folding endurance in the aging experiment were noteworthy (Figure 4c). After 15 days of aging, the folding endurance of UP decreased by 77.78%, from 4.5 times to 1 time, while P1 demonstrated a 49.49% decrease, and P2 exhibited a 42.11% reduction. Compared to the tear index and the tensile index, the folding endurance images of KH550-BC- and KH550-BC/MgO-treated paper intersected between 10 and 15 days, which meant that the folding endurance of the paper was more susceptible to environmental pH and alkali storage [39]. All the above results underscored that the anti-aging property of paper can be effectively enhanced by KH550-BC/MgO.

2.4.2. Resistance to Acidification of Repaired Paper

The pH value and alkali reserve were the important indicators for evaluating the deacidification effect of aged paper samples repaired by KH550-BC and KH550-BC/MgO. The environment of high temperature and humidity accelerated the degradation of the primary components of the paper, resulting in the accumulation of acidic substances [40–42]. This was the main factor contributing to the observed decline in the pH value across all paper samples (Figure 5). In the 30-day artificial aging process, the pH of P1 was gradually changed from 7.96 to 6.65, confirming that the amino groups of KH550-BC exerted a buffer effect on the degradation of paper fibers. However, P2 remained weakly alkaline after 30 days of aging, with a pH value of 7.56 and an alkali reserve of 0.25 mol/kg. This further indicated that the fiber network structure of BC stabilized the nano-MgO and played a positive role in covering [43], thus allowing for a gradual and continuous process of resisting acidification compared to KH550-BC.

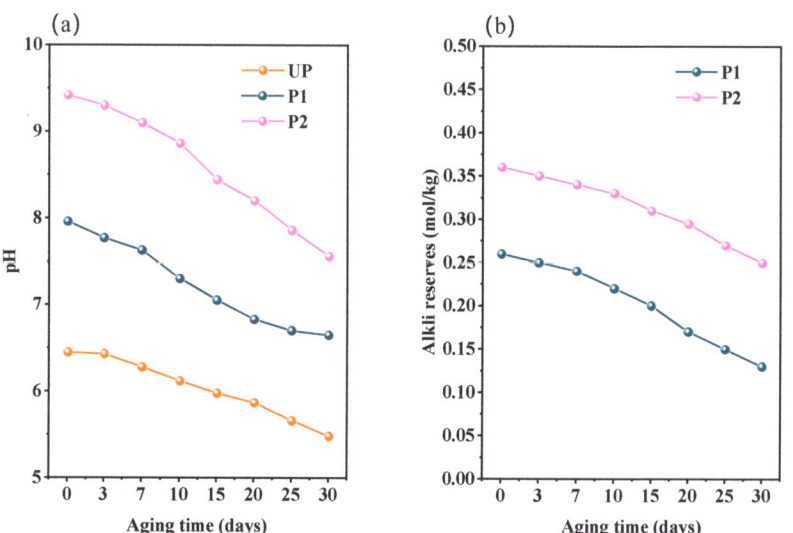

Figure 5. The change curves of (a) the pH and (b) the alkali reserves of paper samples with aging time.

2.4.3. Whiteness and Chromatic Aberration Change during Aging

Consistent with the above changes in mechanical properties and pH properties, aging for 3 days had a negligible impact on the whiteness and color difference of all paper samples (Figure 6). More chromophoric groups were formed during the accelerated lignin degradation in the acidifying UP throughout the hygrothermal aging process which was the cause of the significant yellowing observed in UP [37]. The whiteness decreased from the initial 49.59% to only 34.49%, along with the color difference increased to 8.20 after 30 days of aging. In contrast, the whiteness of P2 treated with KH550-BC/MgO exhibited the slowest decrease, as evidenced by a postponed aging process and a decreased yellowing rate due to the presence of nano-MgO [44]. Aged for 30 days, the values of whiteness and color difference were 41.09% and 4.16, respectively (Figure 6). What is more, the macroscopic scanning images of all of the paper samples (Figure 7) during the aging process corresponded to the change trends of whiteness and chromatic aberration as mentioned earlier, also illustrating that KH550-BC/MgO dispersion was the optimal repair system.

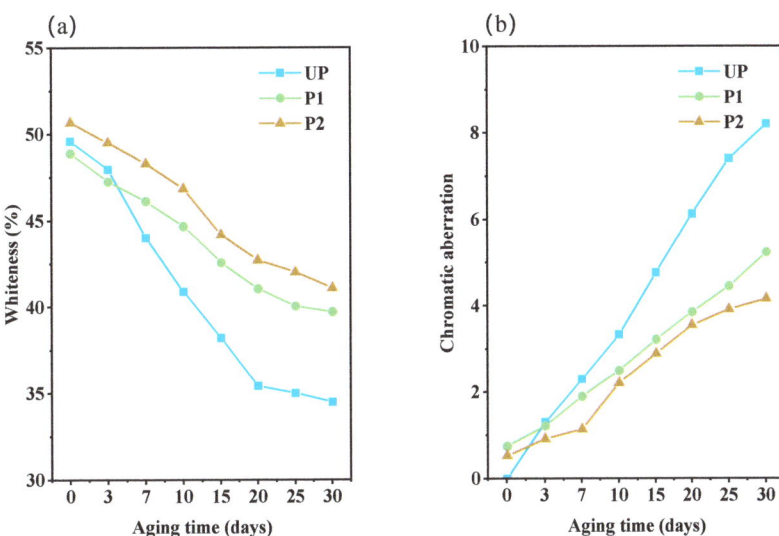

Figure 6. The change curves of (**a**) whiteness and (**b**) chromatic aberration of paper samples with aging time.

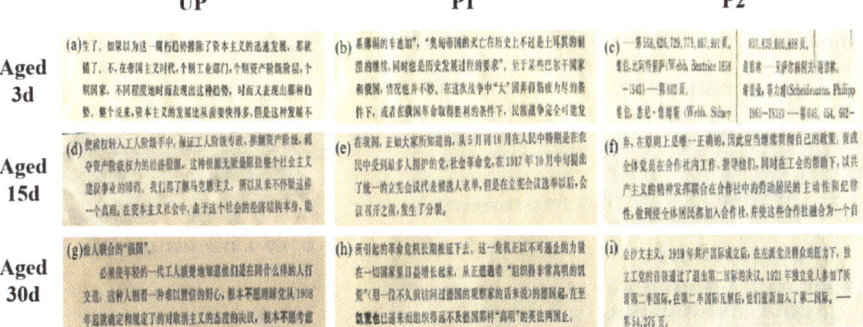

Figure 7. The scanning images of the paper samples with aging time, wherein (**a,d,g**) represented UP, (**b,e,h**) represented P1, and (**c,f,i**) represented P2.

2.5. Anti-Aging Mechanism Analysis

ATR-FTIR spectra were performed on each of the three paper samples to analyze the changes in the functional groups during the aging process [45]. As shown in Figure 8a–c, after repair by KH550-BC and KH550-BC/MgO, the C=O stretching vibration peak of UP at 1650 cm^{-1} was transferred to 1641 cm^{-1} and 1644 cm^{-1}, thereby confirming the existence of hydrogen bonding between the reinforcing agent and the aging paper fiber [46]. New peaks emerged at 1550 cm^{-1} and 1450 cm^{-1}, which were primarily attributed to the bending vibration of NH$_2$ and the stretching vibration of C-N in KH550-BC. This indicated that the modified BC was successfully combined with the paper fibers [24]. Furthermore, in Figure 8c, peaks corresponding to Mg-O stretching vibration and bending vibration were evident at 670 cm^{-1} and 470 cm^{-1}, respectively. The strength of the characteristic peak of MgO decreased throughout the aging process, suggesting that MgO continued to exert an anti-acidifying effect during this period.

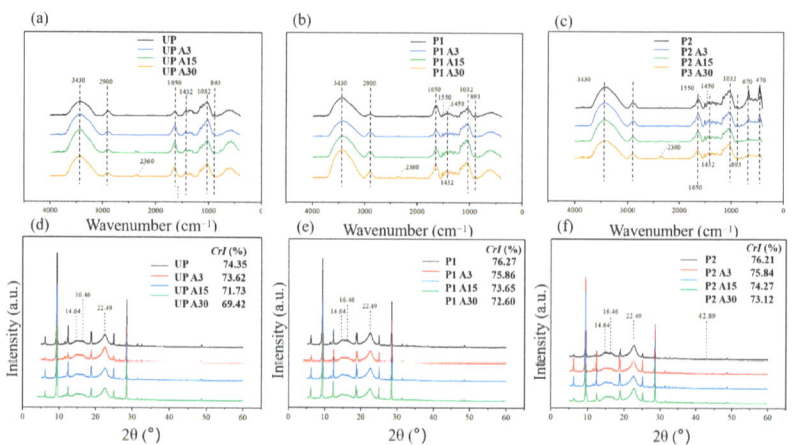

Figure 8. FTIR-transform infrared spectra (**a–c**) and X-ray diffraction patterns (**d–f**) of the paper samples, wherein (**a,d**) represented UP, (**b,e**) represented P1, and (**c,f**) represented P2.

The degree of aging of the different papers can be compard by exploring the crystallinity index of the cellulose in Figure 8d–f [47–49]. Following the application of the two repair systems, the CrI of P2 (76.21%) was marginally higher than that of UP (74.33%) comparable to that of P1 (76.25%). These findings were consistent with the results of the mechanical property analysis (Table 2). With the gradual aging of the paper, the decrease rate of the crystallinity of P1 was lower than that of UP (Figure 8d,e), suggesting that P1 had a certain aging resistance provided by the high strength and favorable interface bonding of KH550-BC. Based on these observations, Figure 8f illustrated that the CrI of P2 exhibited a minimal variation; after aging for 30 d, the CrI was 73.12%, with a reduction of 3.09%. The slowest rate of reduction in the cellulose crystallization index demonstrated the inhibition of P2 aging and the durability of KH550-BC/MgO's protective action [50,51].

SEM analysis was employed to assess the distribution of the active ingredients of the repaired dispersion on paper. After aging for 3 d, the fillers on the surface of UP were observed to have diminished in quantity. However, there were no discernible alterations in the morphology of P1 and P2. The smooth KH550-BC and the rough KH550-BC/MgO were observed to lie on the surface, with the fiber structure remaining apparent (Figure 9a–c). With the extension of aging time (Figure 9d–i), the interfiber fillers of UP decreased noticeably. P1 also showed a reduction in the content of KH550-BC and an increase in the number of pores. However, the fiber morphology of the P2 had the smallest change, the aggregated structures of KH550-BC/MgO could always be recognized to be coated on the fiber surface and bridge between the fibers, and the pore changes were not obvious. It was further noted from Figure 9j,k that the MgO content gradually decreased as it played a role in acid resistance, which was in line with the weakening of its characteristic peak intensity (Figure 8c). The preceding analyses led to the reasonable conclusion that the KH550-BC/MgO is interwoven with the aging paper fiber, forming hydrogen bonds that cooperate with the anti-acidifying effect of MgO to jointly ensure the long-term stable preservation of the repaired paper.

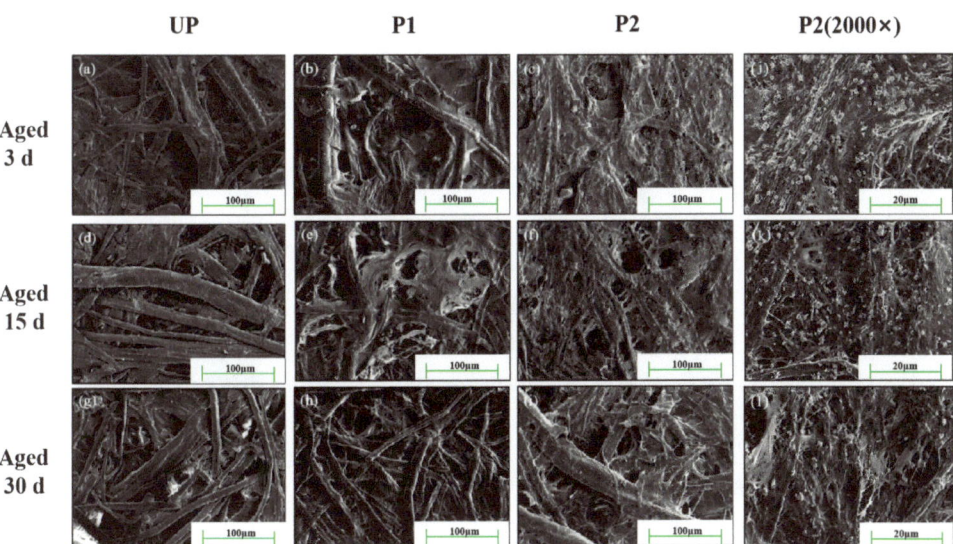

Figure 9. FESEM images of the paper samples during the aging process, wherein (**a–i**) represented 500× magnification and (**j–l**) represented 2000× magnification.

3. Material and Methods

3.1. Material

Bacterial cellulose was supplied by Hainan Guangyu Biotechnology Co., Ltd. (Haikou, China). 3-aminopropyl triethoxysilane (KH550) and nano-MgO were both analytical grade and provided by Shanghai McLean Biochemical Technology Co., Ltd. (Shanghai, China). The paper samples were selected from an old book made of bleached hardwood pulp and published in 1972. After the natural aging process, the pH was found to be 6.45 ± 0.08, with a gram weight of 56.15 g/m^2.

3.2. Preparation of KH550-BC/MgO Dispersion

In total, 1 g dry BC and 5% KH550 solution (pre-hydrolyzed in 160 mL 80% ethanol solution for 30 min) were added to a 500 mL three-mouth flask, reacted at 80 °C and 350 rpm for 4 h. Afterwards, a dispersion of KH550-BC was prepared at 40,000 rpm using a standard disperser (SKG 1246, SKG, Guangzhou, China) and then homogenized at 45 bar 5 times by a high-pressure nanohomogenizer (NanoGenizer, Genizer, Irvine, CA, USA).

A specific mass of nano-MgO was added into a 0.4% concentration of KH550-BC, and the mixture was dispersed uniformly by ultrasonic dispersion (JY 99-IIDN, Scientz, Ningbo, China) for 30 min to obtain a KH550-BC/MgO dispersion. During this experiment, the mass fraction of MgO in the KH550-BC/MgO system was 0.1%, 0.2%, and 0.3%, respectively. The preparation diagram of KH550-BC/MgO is illustrated in Figure 10.

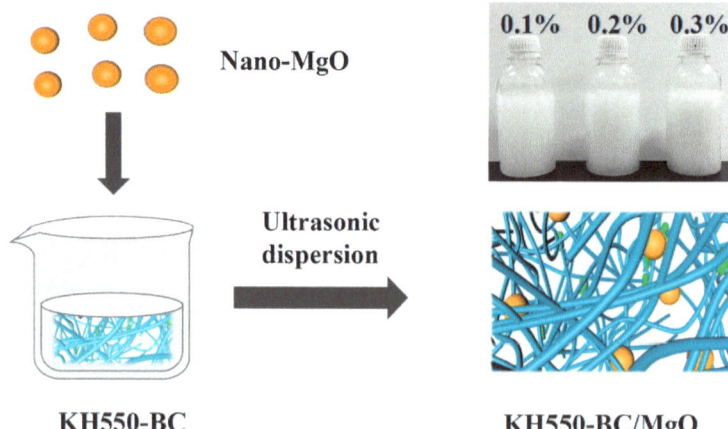

Figure 10. Diagram of the KH550-BC/MgO system preparation.

3.3. Artificial Accelerated Aging of Paper Samples

Aged paper that had no fractures or stains on the surface was selected for this study. Based on our previous reinforcement process [24], the paper samples were individually impregnated for 5 min in KH550-BC and KH550-BC/MgO dispersion at the same concentration (0.4%). Following impregnation, the paper samples were suspended vertically in a vacuum drying oven (BGZ-6050, Shanghai Yiheng Technology Co., Ltd., Shanghai, China) and subjected to a drying process at 50 °C for 0.5 h. They were then equilibrated with water for 24 h, according to ISO 187:1990 [52].

In accordance with ISO 5630-3:1996 [53], the untreated paper (UP), the paper treated with the KH550-BC dispersion (P1), and the paper treated with the KH550-BC/MgO dispersion (P2) were subjected to an artificial hydrothermal ageing test for specific days (3 d, 7 d, 10 d, 15 d, 20 d, 25 d, and 30 d) at 80 °C and 65% relative humidity in a constant temperature and humidity chamber (LHS-100CL, Shanghai Yiheng Technology Co., Ltd., Shanghai, China). Prior to characterization, all paper samples were suspended for 24 h at 23 °C and 50% relative humidity according to ISO 187:1990 [52]. The repair process and aging conditions of the aged paper samples are shown in Figure 11.

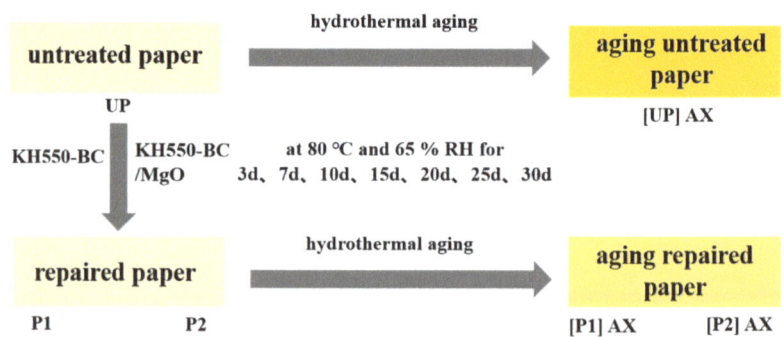

Figure 11. Scheme of the paper samples used and the treatments applied. X is the number of days of the hydrothermal aging test.

3.4. Characterization of Paper Samples

The pH value of the paper samples was determined by cold extraction according to ISO 6588-1:2021 [54]. The alkali reserve of the paper samples was determined by titration according to ISO 10716-1994 [55], and the calculation formula was as follows in Equation (1).

$$X = \frac{(V_2 - V_1)c}{m} \quad (1)$$

where X is the alkali reserve of the paper samples, V_1 and V_2 represent the volume of NaOH consumed by the paper sample solution and blank reagent, respectively, m is the absolute dry mass of the paper samples, and c is the concentration of the NaOH standard solution.

The mechanical properties (tear index, tensile index, folding endurance, and zero-span tensile strength) testing methods had been reported in detail [17]. The tensile stress of the paper samples (4 cm × 8 mm) was determined by a material testing machine (INSTRON 3300, INSTRON, Shanghai, China) at 10 mm/min. In addition, the paper samples were placed on the Surface WCA Tester (OCA40 Micro, Dataphysics, Filderstadt, Germany) to test the surface wettability, and the drip flow of deionized water was controlled to be 8 µL.

The surface morphology of paper samples was obtained by a scanner (Epson Perfection V330, Epson, Beijing, China). In accordance with ISO 11476:2010 [56], a whiteness tester (WSB-2, Xinrui, Shanghai, China) was employed to assess the whiteness, L^*, a^*, and b^* values of the paper samples. The chromatic aberration (ΔE^*) was calculated using the following formula:

$$\Delta E^* = \sqrt{(\Delta L^*)^2 + (\Delta a^*)^2 + (\Delta b^*)^2} \quad (2)$$

where ΔL^* is the difference in lightness and darkness, Δa^* represents the difference in red and green, and Δb^* represents the difference in yellow and blue.

A field emission scanning electron microscope (FESEM; LEO1530VP, Zeiss, Jena, Germany) was used to assess the microscopic morphology of the paper samples. The chemical structure of the paper samples with varying degrees of aging was determined by attenuated total reflection Fourier transform infrared spectroscopy (ATR-FTIR; TENSOR27, Bruker, Ettlingen, Germany). The ATR-FTIR spectrum was captured in absorption mode with a resolution of 4 cm^{-1}, a range of 4000–500 cm^{-1}, and 32 scans. The crystallinity spectrum of the paper samples was obtained by a Bruker X-ray polycrystalline diffractometer (XRD, D8 ADVANCE, Bruker, Ettlingen, Germany) using a conventional BB focusing light path with a 2θ range of 5°~60°, and the crystallinity index was calculated by the Segal formula.

4. Conclusions

This study examined the application of BC-based composites as paper relics reinforcing agents and their potential mechanism of anti-aging. The incorporation of nano-MgO into KH550-BC, which plays an integrated role in deacidification and reinforcement, was found to be an effective method to significantly improve the durability of the paper. The experimental results demonstrate that the paper repaired by KH550-BC/MgO aged for 30 days still exhibits an adequate strength and acid-resistance ability, with approximately 30% higher mechanical properties than untreated paper and an alkali reserve of 0.25 mol/kg. ATR-FTIR and SEM analyses demonstrated that a strong interfacial binding force in coordination with sufficient alkali reserves contributed to these observed properties. These findings offer valuable insights for the long-term stable conservation of ancient books using BC, and the potential antibacterial property of repaired paper given by KH550-BC/MgO should be considered in future studies for the optimal conservation of paper relics.

Author Contributions: Conceptualization, H.M. and X.J.; Methodology, T.W.; Software, X.W.; Validation, H.M. and X.J.; Formal Analysis, H.M.; Investigation, H.M. and H.F.; Resources, H.F.; Data Curation, X.W.; Writing—Original Draft Preparation, H.M. and T.W.; Writing—Review and Editing, H.M. and T.W.; Visualization, X.W.; Supervision, H.Z.; Project Administration, H.Z.; Funding Acquisition, X.J. and H.Z. All authors have read and agreed to the published version of the manuscript.

Funding: This research was funded by the Key Laboratory of Pulp and Paper Science & Technology of the Ministry of Education, Qilu University of Technology (Shandong Academy of Sciences), grant number: KF202113; the National Engineering Lab for Pulp and Paper; and the China National Pulp and Paper Research Institute Co., Ltd., grant number: 20230299.

Institutional Review Board Statement: Not applicable.

Informed Consent Statement: Not applicable.

Data Availability Statement: Data may be shared under request.

Conflicts of Interest: The authors declare that they have no known competing financial interests or personal relationships that could have appeared to influence the work reported in this paper. All the authors listed have approved the manuscript enclosed.

References

1. Zervos, S.; Moropoulou, A. Methodology and Criteria for the Evaluation of Paper Conservation Interventions: A Literature Review. *Restaur.-Int. J. Preserv. Libr. Arch. Mater.* **2006**, *27*, 219–274. [CrossRef]
2. Fan, L.-t.; Gharpuray, M.M.; Lee, Y.-H. Acid Hydrolysis of Cellulose. In *Cellulose Hydrolysis*; Fan, L.-T., Gharpuray, M.M., Lee, Y.-H., Eds.; Springer: Berlin/Heidelberg, Germany, 1987; pp. 121–148.
3. Strlič, M.; Kolar, J.; Pihlar, B. Methodology and analytical techniques in paper stability studies. In *Aging and Stabilisation of Paper*; Strlič, M., Kolar, J., Eds.; University of Ljubljana: Ljubljana, Slovenia, 2005; pp. 27–47.
4. Strlič, M.; Menart, E.; Cigić, I.K.; Kolar, J.; de Bruin, G.; Cassar, M. Emission of reactive oxygen species during degradation of iron gall ink. *Polym. Degrad. Stab.* **2010**, *95*, 66–71. [CrossRef]
5. Croitoru, C.; Roata, I.C. Ionic Liquids as Reconditioning Agents for Paper Artifacts. *Molecules* **2024**, *29*, 963. [CrossRef]
6. Santos, A.; Cerrada, A.; García, S.; San Andrés, M.; Abrusci, C.; Marquina, D. Application of Molecular Techniques to the Elucidation of the Microbial Community Structure of Antique Paintings. *Microb. Ecol.* **2009**, *58*, 692–702. [CrossRef]
7. Magaudda, G. The recovery of biodeteriorated books and archive documents through gamma radiation: Some considerations on the results achieved. *J. Cult. Herit.* **2004**, *5*, 113–118. [CrossRef]
8. Liang, X.; Zheng, L.; Li, S.; Fan, X.; Shen, S.; Hu, D. Electrochemical removal of stains from paper cultural relics based on the electrode system of conductive composite hydrogel and PbO_2. *Sci. Rep.* **2017**, *7*, 8865. [CrossRef]
9. Jiang, F.; Yang, Y.; Weng, J.; Zhang, X. Layer-by-Layer Self-Assembly for Reinforcement of Aged Papers. *Ind. Eng. Chem. Res.* **2016**, *55*, 10544. [CrossRef]
10. Jia, Z.; Yang, C.; Zhao, F.; Chao, X.; Li, Y.; Xing, H. One-Step Reinforcement and Deacidification of Paper Documents: Application of Lewis Base—Chitosan Nanoparticle Coatings and Analytical Characterization. *Coatings* **2020**, *10*, 1226. [CrossRef]
11. Li, S.; Tang, J.; Jiang, L.; Jiao, L. Conservation of aged paper using reduced cellulose nanofibrils/aminopropyltriethoxysilane modified $CaCO_3$ particles coating. *Int. J. Biol. Macromol.* **2023**, *255*, 128254. [CrossRef]
12. Xiang, Z.; Liu, Q.; Chen, Y.; Lu, F. Effects of physical and chemical structures of bacterial cellulose on its enhancement to paper physical properties. *Cellulose* **2017**, *24*, 3513–3523. [CrossRef]
13. Wang, Y.; Luo, W.; Tu, Y.; Zhao, Y. Gelatin-Based Nanocomposite Film with Bacterial Cellulose–MgO Nanoparticles and Its Application in Packaging of Preserved Eggs. *Coatings* **2021**, *11*, 39. [CrossRef]
14. Santos, S.M.; Carbajo, J.M.; Gómez, N.; Quintana, E.; Ladero, M.; Sánchez, A.; Chinga-Carrasco, G.; Villar, J.C. Use of bacterial cellulose in degraded paper restoration. Part I: Application on model papers. *J. Mater. Sci.* **2016**, *51*, 1541–1552. [CrossRef]
15. Gómez, N.; Santos, S.; Carbajo, J.; Villar, J. Use of bacterial cellulose in degraded paper restoration: Effect on visual appearance of printed paper. *Bioresources* **2017**, *12*, 9130–9142. [CrossRef]
16. Chen, X.; Ding, L.; Ma, G.; Yu, H.; Wang, X.; Zhang, N.; Zhong, J. Use of bacterial cellulose in the restoration of creased Chinese Xuan paper. *J. Cult. Herit.* **2023**, *59*, 23–29. [CrossRef]
17. Wu, X.; Mou, H.Y.; Fan, H.; Yin, J.; Liu, Y.; Liu, J. Improving the Flexibility and Durability of Aged Paper with Bacterial Cellulose. *Mater. Today Commun.* **2022**, *32*, 103827. [CrossRef]
18. Nakayama, A.; Kakugo, A.; Gong, J.P.; Osada, Y.; Takai, M.; Erata, T.; Kawano, S. High Mechanical Strength Double-Network Hydrogel with Bacterial Cellulose. *Adv. Funct. Mater.* **2004**, *14*, 1124–1128. [CrossRef]
19. He, H.; Teng, H.; An, F.; Wang, Y.-W.; Qiu, R.; Chen, L.; Song, H. Nanocelluloses review: Preparation, biological properties, safety, and applications in the food field. *Food Front.* **2023**, *4*, 85–99. [CrossRef]
20. Zhang, X.; Yao, J.; Yan, Y.; Huang, X.; Zhang, Y.; Tang, Y.; Yang, Y. Reversible Deacidification and Preventive Conservation of Paper-Based Cultural Relics by Mineralized Bacterial Cellulose. *ACS Appl. Mater. Interfaces* **2024**, *16*, 13091–13102. [CrossRef] [PubMed]
21. Choo, K.W.; Dhital, R.; Mao, L.; Lin, M.; Mustapha, A. Development of polyvinyl alcohol/chitosan/modified bacterial nanocellulose films incorporated with 4-hexylresorcinol for food packaging applications. *Food Packag. Shelf Life* **2021**, *30*, 100769. [CrossRef]
22. Shao, W.; Wu, J.; Liu, H.; Ye, S.; Jiang, L.; Liu, X. Novel bioactive surface functionalization of bacterial cellulose membrane. *Carbohydr. Polym.* **2017**, *178*, 270–276. [CrossRef]
23. Fernandes, S.C.; Sadocco, P.; Alonso-Varona, A.; Palomares, T.; Eceiza, A.; Silvestre, A.J.; Mondragon, I.; Freire, C.S. Bioinspired antimicrobial and biocompatible bacterial cellulose membranes obtained by surface functionalization with aminoalkyl groups. *ACS Appl. Mater. Interfaces* **2013**, *5*, 3290–3297. [CrossRef] [PubMed]
24. Mou, H.; Wu, T.; Wu, X.; Zhang, H.; Ji, X.; Fan, H.; Song, H. Improvement of interface bonding of bacterial cellulose reinforced aged paper by amino-silanization. *Int. J. Biol. Macromol.* **2024**, *275*, 133130. [CrossRef]

25. Ahn, K.; Banik, G.; Potthast, A. Sustainability of Mass-Deacidification. Part II: Evaluation of Alkaline Reserve. *Restaur. Int. J. Preserv. Libr. Arch. Mater.* 2012, *33*, 48–75. [CrossRef]
26. Li, Y.; Wang, J.; Jia, Z.; Zhou, Y.; Chao, X.; Terigele; Li, J.; Li, Y.; Xing, H. Deacidification and consolidation of brittle book paper using bacterial cellulose composite with zinc oxide nanoparticles. *J. Cult. Herit.* 2023, *64*, 83–91. [CrossRef]
27. Jablonsky, M.; Holubkova, S.; Kazikova, J.; Botkova, M.; Haz, A.; Bajzikova, M. The treatment of acid newsprint paper: Evaluation of treatment bymgo or by a mixture of MgO and methyl methoxy magnesium carbonate. *Wood Res.* 2013, *58*, 151–164.
28. Huang, J.; Liang, G.; Lu, G.; Zhang, J. Conservation of acidic papers using a dispersion of oleic acid-modified MgO nanoparticles in a non-polar solvent. *J. Cult. Herit.* 2018, *34*, 61–68. [CrossRef]
29. He, B.; Ai, J.; Qi, S.; Ren, J.; Zhao, L.; Liu, C.; Fan, H. Highly stable nano magnesium oxide organic coatings for nondestructive protection of acidic paper documents. *Prog. Org. Coat.* 2022, *167*, 106833. [CrossRef]
30. Amornkitbamrung, L.; Bračič, D.; Bračič, M.; Hribernik, S.; Malešič, J.; Hirn, U.; Vesel, A.; Stana-Kleinschek, K.; Kargl, R.; Mohan, T. Comparison of Trimethylsilyl Cellulose-Stabilized Carbonate and Hydroxide Nanoparticles for Deacidification and Strengthening of Cellulose-Based Cultural Heritage. *ACS Omega* 2020, *5*, 29243–29256. [CrossRef]
31. Wu, X.; Xiang, Z.; Song, T.; Qi, H. Wet-strength agent improves recyclability of dip-catalyst fabricated from gold nanoparticle-embedded bacterial cellulose and plant fibers. *Cellulose* 2019, *26*, 3375–3386. [CrossRef]
32. Zhang, X.; Yan, Y.; Yao, J.; Jin, S.; Tang, Y. Chemistry directs the conservation of paper cultural relics. *Polym. Degrad. Stab.* 2023, *207*, 110228. [CrossRef]
33. Kumar, S.; Chauhan, V.S.; Chakrabarti, S.K. Separation and analysis techniques for bound and unbound alkyl ketene dimer (AKD) in paper: A review. *Arab. J. Chem.* 2016, *9*, S1636–S1642. [CrossRef]
34. Basta, A.; El-Saied, H.; Mohamed, S.; El-sherbiny, S. The Role of Neutral Rosin-Alum Size in the Production of Permanent Paper. *Restaur.-Int. J. Preserv. Libr. Arch. Mater.* 2006, *27*, 67–80. [CrossRef]
35. Wu, C.; Liu, Y.; Hu, Y.; Ding, M.; Cui, X.; Liu, Y.; Liu, P.; Zhang, H.; Yang, Y.; Zhang, H. An Investigation into the Performance and Mechanisms of Soymilk-Sized Handmade Xuan Paper at Different Concentrations of Soymilk. *Molecules* 2023, *28*, 6791. [CrossRef]
36. Kono, H.; Uno, T.; Tsujisaki, H.; Anai, H.; Kishimoto, R.; Matsushima, T.; Tajima, K. Nanofibrillated Bacterial Cellulose Surface Modified with Methyltrimethoxysilane for Fiber-Reinforced Composites. *ACS Appl. Nano Mater.* 2020, *3*, 8232–8241. [CrossRef]
37. Zervos, S. Natural and Accelerated Ageing of Cellulose and Paper: A Literature Review. In *Cellulose Structure and Properties, Derivatives and Industrial Uses*; Lejeune, A., Deprez, T., Eds.; Nova Science Publishers: Happauge, NY, USA, 2010; pp. 155–203.
38. Wong, K.K.Y.; Richardson, J.D.; Mansfield, S.D. Enzymatic Treatment of Mechanical Pulp Fibers for Improving Papermaking Properties. *Biotechnol. Prog.* 2000, *16*, 1025–1029. [CrossRef]
39. Liu, J.; Gao, S.; Wang, Y.; Li, X.; Xuan, C. Effect of ambient humidity on water content and mechanical properties of paper. *J. Beijing Inst. Print.* 2013, *21*, 5–8.
40. He, B.; Lin, Q.; Chang, M.; Liu, C.; Fan, H.; Ren, J. A New and Highly Efficient Conservation Treatment for Deacidification and Strengthening of Aging Paper by In-situ Quaternization. *Carbohydr. Polym.* 2019, *209*, 250–257. [CrossRef]
41. Zou, X.; Uesaka, T.; Gurnagul, N. Prediction of paper permanence by accelerated aging I. Kinetic analysis of the aging process. *Cellulose* 1996, *3*, 243–267. [CrossRef]
42. Łojewski, T.; Miśkowiec, P.; Molenda, M.; Lubańska, A.; Łojewska, J. Artificial versus natural ageing of paper. Water role in degradation mechanisms. *Appl. Phys. A* 2010, *100*, 625–633. [CrossRef]
43. Safaei, M.; Taran, M. Preparation of Bacterial Cellulose Fungicide Nanocomposite Incorporated with MgO Nanoparticles. *J. Polym. Environ.* 2022, *30*, 2066–2076. [CrossRef]
44. Liang, G.; Lu, G.; Zhang, J.; Zhen, D. Study on the application of nano-magnesium oxide to the deacidification of paper cultural relics. *New Chem. Mater.* 2017, *45*, 244–246.
45. Hajji, L.; Boukir, A.; Assouik, J.; Pessanha, S.; Figueirinhas, J.L.; Carvalho, M.L. Artificial aging paper to assess long-term effects of conservative treatment. Monitoring by infrared spectroscopy (ATR-FTIR), X-ray diffraction (XRD), and energy dispersive X-ray fluorescence (EDXRF). *Microchem. J.* 2016, *124*, 646–656. [CrossRef]
46. Wang, Y.J.; Zhang, X.N.; Song, Y.; Zhao, Y.; Chen, L.; Su, F.; Li, L.; Wu, Z.L.; Zheng, Q. Ultrastiff and Tough Supramolecular Hydrogels with a Dense and Robust Hydrogen Bond Network. *Chem. Mater.* 2019, *31*, 1430–1440. [CrossRef]
47. Jinquan, W.; Yongwen, M.; Yan, W.; Chen, Y. The Content of Different Hydrogen Bond Models and Crystal Structure of Eucalyptus Fibers during Beating. *BioResources* 2013, *8*, 717–734.
48. Wu, C.; Jin, C.; Zhu, Z.; Liu, P.; Zhang, H. Discussion on the method of detecting the crystalline structure of cellulose in paper. *Fudan J. (Nat. Sci. Ed.)* 2022, *61*, 589–597.
49. Ju, X.; Bowden, M.; Brown, E.E.; Zhang, X. An improved X-ray diffraction method for cellulose crystallinity measurement. *Carbohydr. Polym.* 2015, *123*, 476–481. [CrossRef]
50. Hassan, R. Thermal degradation of paper: The structural changes of fibres. *Egypt. J. Archaeol. Restor. Stud.* 2016, *6*, 71–84.
51. Ma, X.; Tian, S.; Li, X.; Fan, H.; Fu, S. Combined Polyhexamethylene Quanidine and Nanocellulose for the Conservation and Enhancement of Ancient Paper. *Cellulose* 2021, *28*, 8027–8042. [CrossRef]
52. *ISO 187:1990*; Paper, Board and Pulps—Standard Atmosphere for Conditioning and Testing and Procedure for Monitoring the Atmosphere and Conditioning of Samples. ISO: Geneva, Switzerland, 1990. Available online: https://www.iso.org/standard/4037.html (accessed on 16 August 2024).

53. *ISO 5630-3:1996*; Paper and Board—Accelerated Ageing Part 3: Moist Heat Treatment at 80 Degrees C and 65% Relative Humidity. ISO: Geneva, Switzerland, 1996. Available online: https://www.iso.org/standard/23402.html (accessed on 16 August 2024).
54. *ISO 6588–1:2021*; Paper, Board and Pulps—Determination of pH of Aqueous Extracts—Part 1: Cold Extraction. ISO: Geneva, Switzerland, 2021. Available online: https://www.iso.org/standard/83250.html (accessed on 16 August 2024).
55. *ISO 10716-1994*; Paper and Board—Determination of Alkali Reserve. ISO: Geneva, Switzerland, 1994. Available online: https://www.iso.org/standard/18809.html (accessed on 16 August 2024).
56. *ISO 11476:2010*; Paper and Board—Determination of CIE Whiteness, C/2 Degrees (Indoor Illumination Conditions). ISO: Geneva, Switzerland, 2010. Available online: https://www.iso.org/standard/44136.html (accessed on 16 August 2024).

Disclaimer/Publisher's Note: The statements, opinions and data contained in all publications are solely those of the individual author(s) and contributor(s) and not of MDPI and/or the editor(s). MDPI and/or the editor(s) disclaim responsibility for any injury to people or property resulting from any ideas, methods, instructions or products referred to in the content.

MDPI AG
Grosspeteranlage 5
4052 Basel
Switzerland
Tel.: +41 61 683 77 34

Molecules Editorial Office
E-mail: molecules@mdpi.com
www.mdpi.com/journal/molecules

Disclaimer/Publisher's Note: The title and front matter of this reprint are at the discretion of the Guest Editors. The publisher is not responsible for their content or any associated concerns. The statements, opinions and data contained in all individual articles are solely those of the individual Editors and contributors and not of MDPI. MDPI disclaims responsibility for any injury to people or property resulting from any ideas, methods, instructions or products referred to in the content.

www.ingramcontent.com/pod-product-compliance
Lightning Source LLC
LaVergne TN
LVHW072357090526
838202LV00019B/2566